Image Analysis for the Biological Sciences

Image Analysis for the Biological Sciences

C. A. Glasbey and G. W. Horgan
Scottish Agricultural Statistics Service
University of Edinburgh

JOHN WILEY & SONS

Chichester · New York · Brisbane · Toronto · Singapore

Copyright © 1995 by John Wiley & Sons Ltd,
Baffins Lane, Chichester,
West Sussex PO19 1UD, England

National   Chichester (01243) 779 777
International (+44) 1243 779 777

Fig. 1.3 Image reproduced by permission of
the Musée du Louvre, Paris

Other Wiley Editorial Offices

John Wiley & Sons, Inc., 605 Third Avenue,
New York, NY 10158–0012, USA

Jacaranda Wiley Ltd, 33 Park Road, Milton,
Queensland 4064, Australia

John Wiley & Sons (Canada) Ltd, 22 Worcester Road,
Rexdale, Ontario M9W 1L1, Canada

John Wiley & Sons (SEA) Pte Ltd, 37 Jalan Pemimpin # 05–04
Block B, Union Industrial Building, Singapore 2057

QH
324.2
G58
1995

Library of Congress Cataloging-in-Publication Data

Glasbey, C. A. (Chris A.)
    Image analysis for the biological sciences / C. A. Glasbey and G. W.
Horgan.
        p.   cm. — (statistics in practice)
    Includes bibliographical references (p.   ) and index.
    ISBN 0 471 93726 6
    1. Biology—Data processing.   2. Biology—Statistical methods.
    3. Image processing—Digital techniques.   4. Image processing—
Statistical methods. I. Horgan, G. W. (Graham W.)   II. Title.
    III. Series: Statistics in practice (Chichester, England)
    QH324.2.G58   1994   94–18579
    574'.0285—dc20   CIP

British Library Cataloguing in Publication Data

A catalogue record for this book is available from the British Library

ISBN   0 471 93726 6

Typeset in 10/12pt Photina by Thomson Press (India) Ltd, New Delhi
Printed and bound in Great Britain by Biddles Ltd, Guildford and King's Lynn

To
our families

Contents

Preface

Quantitative image analysis is the extraction of information from data that are in the form of pictures. The aim of this book is to cover the basics of image analysis from a statistical perspective, and with emphasis on methods relevant to the biological sciences. The book is written for both biological scientists and applied statisticians, who, we assume, are mainly interested in analysing images of static objects in laboratory-type situations. No greater knowledge of mathematics, statistics or computing is necessary than would be gained in a first degree in a biological subject.

Image analysis methods are presented in five self-contained chapters. Detailed algorithms are given for the most important techniques, for the reader who wants to write his/her own computer program. Otherwise, the understanding gained from reading this book can guide the scientist in making intelligent use of the many computer packages on the market.

Fourteen types of image, drawn from a range of applications in microscopy, medical scanning systems, remote sensing, electrophoresis and photography, are used to motivate and illustrate the methods considered. All images, except the Landsat one, are available by anonymous FTP (file transfer) from Internet site peipa.essex.ac.uk. In case of difficulties, contact the publishers.

We are grateful to our collaborators for permission to use their data. They are as follows:

Algal cells	Dr N. J. Martin	Scottish Agricultural College
Cashmere fibres	Dr A. J. F. Russel	Macaulay Land Use Research Institute
DNA sequencing gel	Dr F. G. Wright	Scottish Agricultural Statistics Service
Electrophoretograms	Prof. D. Walliker	University of Edinburgh
Fish	Dr N. J. C. Strachan	Torry Research Station
Fungal hyphae	Dr K. Ritz	Scottish Crop Research Institute
Landsat	Dr C. H. Osman	Macaulay Land Use Research Institute
Magnetic resonance images	Dr M. A. Foster	University of Aberdeen
Muscle fibres	Dr C. A. Maltin	Rowett Research Institute
Synthetic aperture radar	Dr E. Attema	European Space Agency
Soil aggregate	Dr J. F. Darbyshire	Macaulay Land Use Research Institute
Turbinate bones	Dr J. F. Robertson	Scottish Agricultural College
Ultrasound image and X-ray CT	Dr G. Simm	Scottish Agricultural College

The Landsat image is © National Remote Sensing Centre Ltd, Farnborough, Hampshire. We also thank Jim Young, Department of Geography, Edinburgh University, for initially making the data available to us.

We are indebted to our friends and colleagues who commented on earlier versions of the book. We gratefully acknowledge the advice and encouragement of Ian Craw, John Darbyshire, David Elston, Paul Fowler, Sheila Glasbey, Alison Gray, David Hitchcock, John Marchant, Geoff Simm, Norval Strachan, and Mark Berman, along with his colleagues Leanne Bischof, Ed Breen, Yuchong Jiang, Guy Peden and Changming Sun in the CSIRO Division of Mathematics and Statistics.

We also thank Vic Barnett, the Series Editor, for the challenge and guidance in writing this book, and Helen Ramsey and David Ireland of John Wiley and Sons for their technical help. Most of the figures in this book were produced using Zimage, written by the CSIRO Division of Mathematics and Statistics, Sydney.

Finally, we appreciate the backing of Rob Kempton, Director of the Scottish Agricultural Statistics Service, and acknowledge the financial support of the Scottish Office Agriculture and Fisheries Department.

Chris Glasbey
Graham Horgan
Edinburgh
February 1994

Series Preface

Statistics in Practice is an important international series of texts that provide direct coverage of statistical concepts, methods and worked case studies in specific fields of investigation and study.

With sound motivation and many worked practical examples, the books show in down-to-earth terms how to select and use a specific range of statistical techniques in a particular practical field within each title's special topic area.

The books meet the need for statistical support required by professionals and research workers across a wide range of employment fields and research environments. The series covers a wide variety of subject areas: in medicine and pharmaceutics (e.g. in laboratory testing or clinical trials analysis); in social and administrative fields (e.g. for sample surveys or data analysis); in industry, finance and commerce (e.g. for design or forecasting); in the public services (e.g. in forensic science); in the earth and environmental sciences; and so on.

But the books in the series have an even wider relevance than this. Increasingly, statistical departments in universities and colleges are realizing the need to provide at least a proportion of their course-work in highly specific areas of study to equip their graduates for the work environment. For example, it is common for courses to be given on statistics applied to medicine, industry, social and administrative affairs, and the books in this series provide support for such courses.

It is our aim to present judiciously chosen and well-written workbooks to meet everyday practical needs. Feedback of views from readers will be most valuable to monitor the success of this aim.

Vic Barnett
Series Editor
1994

1

Introduction

Image analysis is simply the *extraction of information from pictures*. For example, it's what you are doing in reading these words. Sight is (arguably) the most important sense we have. It is said that 'a picture is worth a thousand words' (a view shared by Charlie Brown's little sister, Sally—see Fig. 1.1). This book is about images that arise in the biological sciences, and analysing them using a computer. In the following three sections we shall address three questions:

- Why use a computer to analyse images?
- What are the data to be analysed?
- What does image analysis consist of?

Finally, in §1.4 we shall summarize the points covered.

1.1 WHY USE A COMPUTER?

If we are so effective at seeing then why try to use a computer to analyse images? Put simply, the answer is that some image analysis is more easily done by the human eye, but for other tasks a computer is better. To illustrate the point, let us consider a problem that arose in a veterinary investigation.

1.1.1 An example of image analysis

To study atrophic rhinitis, a disease of the upper respiratory tract in pigs, snouts of dead pigs were cut in half (Robertson, Wilson and Smith, 1990). The exposed cross-sections were then inked and printed on paper. Figure 1.2(a) shows a print from a disease-free pig. The disease affects the turbinate bones, the curled structures which can be seen, printed as black, inside each of the nasal cavities. Done *et al.* (1984) proposed a morphometric index to measure deterioration in these bones. It is calculated as the ratio of air-space area in the cross-section of the nasal cavities to air-space area plus turbinate-bone area. In order to measure these areas, Robertson *et al.* enhanced the prints labor-

1

Fig. 1.1 Peanuts cartoon. (Sally: © 1960 United Feature Syndicate, Inc. Reprinted by permission.)

iously by hand, using a black pen to fill in areas of bone that had not been printed clearly and typists' correction fluid to whiten places where ink had got by mistake. Figure 1.2(b) shows the result obtained from Fig. 1.2(a). The pictures were then converted into digital form by scanning them into a simple computer-based image analysis system. (See Appendix for more details about computer equipment.) Areas of cavity and turbinate bone were measured, following a further manual operation using a computer mouse to outline the appropriate regions. The whole process had to be repeated on 1200 snouts, which proved to be an exceedingly tedious task and took 300 hours.

This combination of human and computer interpretation is known as *semi-automatic* image analysis. It would be very difficult, maybe even impossible, to program a computer to achieve the enhancement shown in Fig. 1.2(b). It would also be very laborious for a scientist to accurately measure the areas. However, it might have been possible to have made more use of the computer in image enhancement, and thereby reduce the human input. (Later, in Chapter 5, we shall consider some other methods for cleaning-up the turbinate image.)

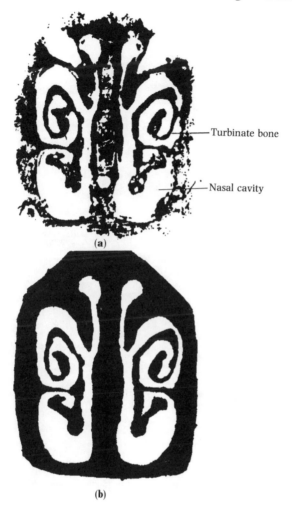

Fig. 1.2 Turbinate image: (a) as printed; (b) after manual enhancement.

Let us return to a general consideration of the relative benefits of human and computer interpretation of images.

1.1.2 Human and computer interpretation of images

The human vision system is superb, particularly at qualitative tasks. For example, look at Fig. 1.3. You may need to half-close your eyes, or hold the book at arm's length, to recognize the picture. It is remarkable that we can see so much even though the picture is so coarse. (If you don't recognize the picture, we shall tell you in two paragraphs' time.)

Fig. 1.3   Example of image with coarse resolution. What is it?

Our eyes can, however, be fooled. In Fig. 1.4(a), both central circles are the same size, although the right-hand one looks bigger. In Fig. 1.4(b), the central squares are the same shade of grey, although the right-hand one looks lighter. Therefore, for extracting quantitative information from images, a computer may be better: it may be more consistent from day to day and not rely on the opinions of different observers. Furthermore, computers may spare us from much tedious image interpretation such as that already discussed for the turbinate images.

Most of the time we see effortlessly. When scientists first started programming computers to interpret images, they expected to make rapid progress. It has proved, however, to be a very difficult task. Perhaps this is because we are

Table 1.1 Data in Fig. 1.3. Each number represents the brightness in the display of a single box in Fig. 1.3, ranging from 0 for black to 255 for white.

132	139	144	146	142	134	151	153	167	166	136	49	26	21	17	18	24	24	43	39	17	4	2	11	8	3	5	9	7
128	136	135	130	124	132	139	135	168	170	111	39	24	16	14	19	30	40	40	34	16	8	9	3	9	0	5	3	4
126	127	125	124	138	158	157	128	93	83	46	18	11	10	11	17	18	25	23	9	5	6	3	5	6	7	10	13	8
123	125	119	125	130	148	117	46	17	16	15	11	14	14	11	12	11	9	6	5	4	6	6	7	23	64	45	16	8
137	129	124	132	119	47	22	39	63	34	48	83	53	21	11	8	12	9	10	7	6	8	7	21	63	147	160	24	9
129	134	135	89	15	92	162	174	169	63	69	174	175	110	69	43	6	2	2	3	5	15	74	127	162	182	190	63	12
130	143	121	17	46	208	242	207	196	77	57	150	153	126	102	126	73	2	0	2	6	65	144	176	185	173	149	93	17
145	147	62	19	82	227	248	210	211	140	124	128	134	47	42	69	165	23	6	6	36	127	170	190	191	190	200	187	19
158	150	29	11	54	181	225	205	168	28	55	103	98	62	39	49	79	16	20	77	85	135	175	169	186	221	253	223	26
170	146	16	8	25	111	153	136	137	48	86	199	172	142	74	77	37	10	8	39	82	108	149	166	181	212	255	237	41
172	157	20	6	6	33	62	89	121	31	52	194	146	85	71	46	15	9	10	25	68	120	162	182	188	205	239	209	42
161	157	47	9	6	9	19	52	88	25	112	108	59	33	28	17	12	10	16	31	75	131	153	158	156	181	210	124	24
153	158	89	17	11	6	8	17	21	17	34	23	13	10	9	7	10	10	18	36	72	101	110	118	117	110	115	36	11
158	166	151	38	17	14	8	7	9	10	11	12	14	13	13	14	12	11	10	11	27	47	48	44	38	33	46	17	5
162	171	162	134	33	16	11	6	7	9	12	13	13	12	12	13	9	4	2	5	6	11	10	11	19	30	4	0	
172	163	158	164	155	70	13	6	5	8	6	10	12	11	12	11	10	8	3	2	7	11	12	11	12	42	25	3	1
159	154	152	155	173	180	159	70	34	22	15	14	12	11	11	12	10	6	5	5	8	7	3	2	25	57	9	2	7
152	144	137	143	156	178	152	108	57	36	30	34	46	15	18	19	16	10	7	2	0	4	5	5	33	50	5	1	4

(a)

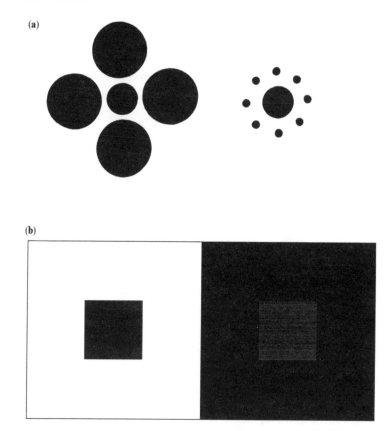

(b)

Fig. 1.4   Two optical illusions: (a) Which central black disc is the larger? (b) Which central grey square is the darker?

not conscious of what processes we go through in looking at, for example, Fig. 1.3. To a computer, an image is no more than a matrix of numbers. Look at Table 1.1, which is the numerical version of Fig. 1.3, a coarse-resolution approximation to Leonardo da Vinci's painting of the Mona Lisa. Table 1.1 is a 29 × 18 array of numbers, ranging in value from 0 to 255, while Fig. 1.3 is a 29 × 18 array of square blocks of varying shades of grey. The values in Table 1.1 are matched to the grey levels in Fig. 1.3, in that 0 is displayed as black, 255 as white, and in general the higher the value the lighter the shading of the corresponding block. Figure 1.5 gives an alternative graphical representation of these numbers, where height is used to represent numerical value. In either case, in looking at Table 1.1 or at Fig. 1.5, we are no better off than the computer—we also have great difficulty in recognizing the Mona Lisa.

Biological objects tend to be more irregular and variable in shape than artificial ones. Therefore they present an even greater challenge to fully automatic

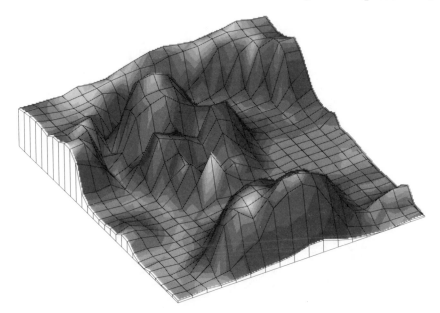

Fig. 1.5   Alternative display of Fig 1.3, with grey level replaced by height.

interpretation. For most practical applications in the biological sciences, the best that can be achieved at present is a semi-automatic system, with the computer reducing the tedious aspects of human image interpretation and carrying out the more quantitative tasks. Fortunately, we are not trying to build a robot or some other autonomous system. If a computer has to recognize an airport runway in order to land a plane then it had better be right *every single time*! In applications involving medical diagnosis, it is usually better for a computer screening system to err on the cautious side, and flag the likely presence of a tumour (for example) if in doubt, in preference to ever missing a real tumour. In the applications we are going to consider, occasional mistakes can be tolerated. In a semi-automatic system, they can either be corrected manually or they can be allowed to remain, in which case they will contribute to the overall uncertainty in the results.

So, what are these applications?

1.1.3   Further examples

In the Scottish Agricultural Statistics Service, which provides statistical and mathematical support to agricultural and environmental research scientists, we encounter many applications of image analysis. The objects that are imaged range in scale from the microscopic to satellite views of the earth.

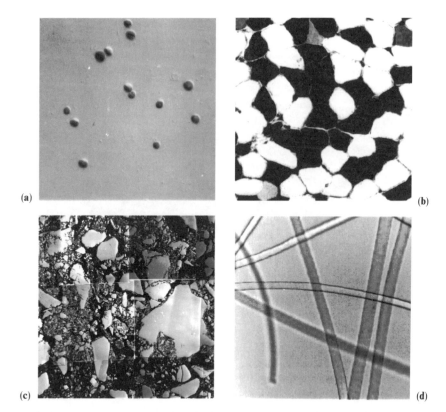

(a)   (b)

(c)   (d)

Fig. 1.6   Microscope images: (a) algal cells; (b) muscle fibres; (c) soil aggregate; (d) cashmere fibres.

Here we will introduce some of these examples, and the questions that the scientists wish to see answered. Throughout the book we will use these images and questions to motivate the methods we consider, and to illustrate the results.

Figure 1.6 shows four *microscope images*. Such digital images would normally be produced by mounting a digitizing camera (see Appendix) on top of a microscope. Those shown here were produced less directly by taking photographs and using a desktop scanner to digitize them (Glasbey, Horgan and Hitchcock, 1994).

- *Algal cells*   Figure 1.6(a) is a differential interference contrast (DIC) microscope image of some algal cells, collected as part of a research programme to manage algal ponds for waste treatment (Martin and Fallowfield, 1989). The aim is to identify, count and measure cells in such images. Note that the effect of DIC is to make one side of each cell light and the other side dark, producing an effect similar to the illumination of 3D objects. DIC microscopy operates by splitting a beam of light, directing one half through the specimen while the other half bypasses it, then combining them back together. When the two beams of light are in phase, they produce after combination a bright area in the image, whereas when they are out of phase they cancel out and produce a dark area (Holmes and Levy, 1987, 1988). DIC is particularly effective in viewing specimens that are almost transparent, because such specimens still change the phase of light passing through them.

- *Muscle fibres*   Figure 1.6(b) is a section through one of the muscles in a rat's leg, the soleus muscle. The transverse section has been stained to demonstrate the activity of Ca^{2+}-activated myofibrillar ATPase and allows one to classify three types of fibre: fast-twitch oxidative glycolytic (dark), slow-twitch oxidative (light) and fast-twitch glycolytic (mid-grey) (Maltin *et al.*, 1989). Information of the numbers and sizes of the fibres are required for research into clenbuterol, a drug that enhances muscle development.
- *Soil aggregate*   Figure 1.6(c) is a montage of backscattered electron scanning micrographs of a soil aggregate embedded in acrylic resin (Darbyshire *et al.*, 1989; Glasbey, Horgan and Darbyshire, 1991). The black areas are soil pores and the lighter areas are the inorganic and organic soil matrix. Note that the image is made more difficult to interpret by the variation in brightness between micrographs in the montage and by the visible edges of some prints. The objective was to study porosity and pore-size distribution within a sample of soil aggregates and to relate these characteristics to microbial activity within and outside the aggregates.
- *Cashmere fibres*   Figure 1.6(d) is a back-illuminated image of cashmere goat fibres. The aim was to measure their diameters automatically, in support of a goat breeding programme (Russel, 1991). Measurement is made more difficult by some fibres being out-of-focus, which gives either dark or light edges to the fibres, so-called 'Becke lines'.

Figure 1.7 shows four digital images produced by *medical scanning systems*:

- *Magnetic resonance imaging (MRI)*   Figures 1.7(a) and (b) are inversion recovery and proton density images through a supine woman's chest. The woman has a cubic test object between her breasts. Magnetic resonance imaging is a non-invasive technique for viewing sections through living tissue by means of large, oscillating magnetic fields. These images were obtained as part of a study of changes in breast volume during the menstrual cycle (Fowler *et al.*, 1990).
- *X-ray computer tomography (CT)*   Figure 1.7(c) is a cross-section through the thorax of a live sheep, obtained in order to estimate the quantity of fat and lean tissue (Simm, 1992). In CT, X-rays are projected through a subject from different directions, and a computer reconstructs an image of the distribution of tissue types from the transmitted X-rays. As is the convention with X-ray plates, light areas in the image denote regions that transmitted

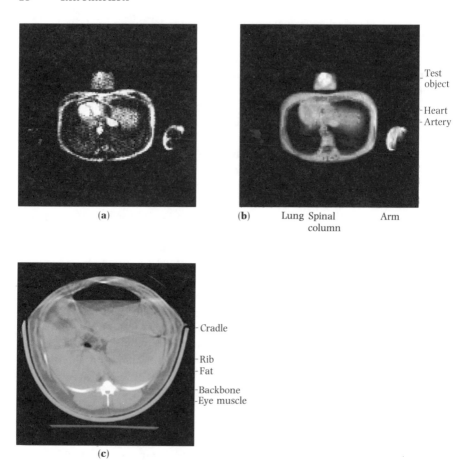

(a)

(b)   Lung Spinal   Arm
           column

Test
object

Heart
Artery

(c)

Cradle

Rib
Fat

Backbone
Eye muscle

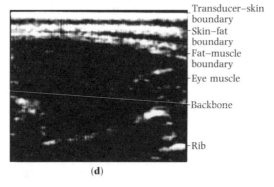

Transducer–skin
boundary
Skin–fat
boundary
Fat–muscle
boundary
Eye muscle

Backbone

Rib

(d)

Fig. 1.7 Medical scanning images: (a) MRI inversion recovery image; (b) MRI proton density image; (c) X-ray CT image; (d) ultrasound image.

less X-rays. The lightest areas are the backbone and the parts of the ribs that intersect the imaging plane. The muscles and internal organs appear slightly lighter than the fat tissue because they are slightly more opaque to X-rays. The U-shaped plastic cradle in which the sheep was lying can also be seen. X-ray attenuation is measured in Hounsfield units, which range between -1000 and about 1000.

- *Ultrasound imaging* Figure 1.7(d) is an ultrasound image of a cross-section through a sheep's back. The instrument operates by using a transducer to send a pulse of sound waves of very high frequency into a subject. When the ultrasound wave meets a boundary between two tissues, partial reflection occurs. The reflected energy is received by the transducer and converted into electrical signals, which are displayed on a video monitor, with time delay interpreted as depth. Unfortunately, in this application, it proved impossible to extract the data directly from the ultrasound machine as there was no video output. Therefore the video display was photographed and redigitized, resulting in the vertical display lines that can be seen in Fig. 1.7(d). Images were collected for the same reason as the CT image above, namely to estimate sheep body composition (Simm, 1992). The topmost approximately horizontal white line is the transducer–skin boundary, below which are the skin–fat and fat–muscle boundaries. The backbone is on the bottom left, from which a rib can be seen sloping slightly upwards. Ultrasound images are far less clear than X-ray images, but have the advantages that the machines are safer, cheaper and more portable.

Figure 1.8 shows some *remotely-sensed images*:

- *Landsat Thematic Mapper (TM) images* Figures 1.8(a–f) are TM bands 1–5 and 7, for a region between the river Tay and the town of St Andrews on the east coast of Scotland, in May 1987. Spatial resolution is 30 m, and the bands correspond to blue (0.45–0.52 μm), green (0.52–0.60 μm), red (0.63–0.69 μm) and three near-infrared (0.76–0.90, 1.55–1.75 and 2.08–2.35 μm) wavelength regions of light reflected from the earth's surface. Many features can be identified. For example, the dark areas at the top and right of Fig. 1.8(f) are water, the bright areas to the centre and left of Fig. 1.8(b) are fields of oil-seed rape, and the section of land in the top-left corner is part of the city of Dundee.
- *Synthetic aperture radar (SAR) image* Figure 1.8(g) is a C-band, HH-polarization, SAR image of an area near Thetford Forest, England, in August 1989. It was obtained by plane as part of the Maestro-1 campaign (Joint Research Centre, Ispra, report IRSA/MWT/4.90). The resolution is 12 m, formed by averaging four adjacent values in the original 3 m × 12 m data. Notice that the intensity values in this image are locally much more variable than in the Landsat images. The apparently random fluctuations are called 'speckle'. Horgan (1994) found good agreement between the observed speckle variance and a theoretically expected value of 0.41, after the data

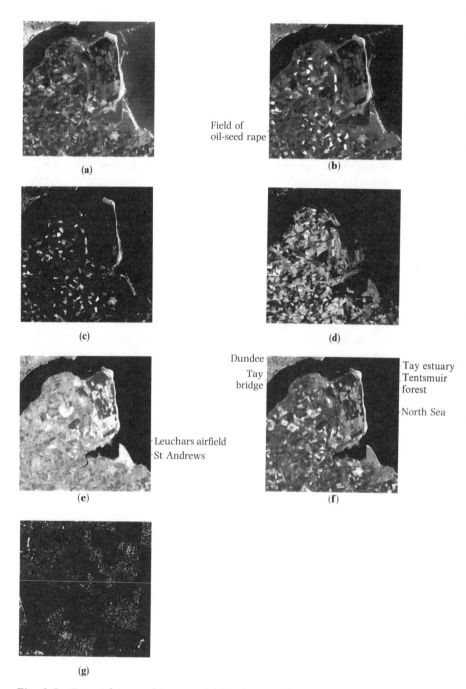

Fig. 1.8   Remotely-sensed images: (a) Landsat TM band 1; (b) band 2; (c) band 3; (d) band 4; (e) band 5; (f) band 7; (g) SAR image.

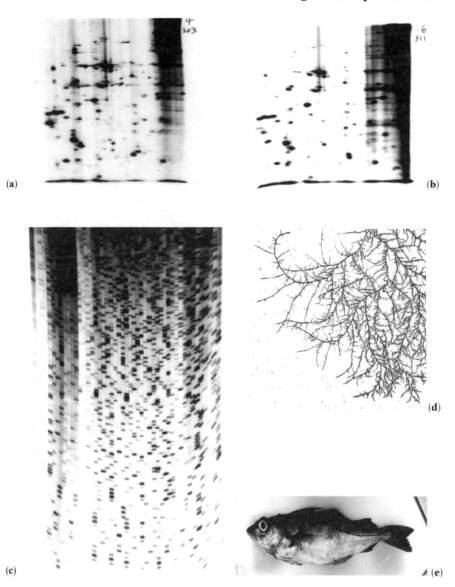

Fig. 1.9   Further illustrative images: (a) 2D electrophoretogram of strain of malaria parasite; (b) electrophoretogram of second strain of malaria parasite; (c) DNA-sequencing gel autoradiograph; (d) fungal hyphae, (e) a fish.

have been log-transformed (see §2.2). Unlike Landsat, SAR is an active sensing system: microwave radiation is beamed down to the earth's surface, a sensor detects the reflected signal, and from this an image is constructed. This sensing system can operate during day or night and is unaffected by clouds unlike Landsat images.

Finally, Fig. 1.9 shows five images obtained directly by digitizing views of the objects themselves:

- *Electrophoretograms*   Figures 1.9(a, b) are digitized autoradiographs of SDS–PAGE gel electrophoretograms of two strains of the malaria parasite *Plasmodium falciparum*. Each spot on a gel represents a different protein. Relative locations of spots, identified by making comparisons between gels, are used to identify the malarial strain (Horgan, Creasey and Fenton, 1992). Initially, a mixture of radioactively labelled proteins was positioned in the top left-hand corner of each gel, from where the individual proteins migrated at different speeds across and down the gel. Then a photographic plate was placed over the gel. This blackened at protein locations in response to radioactive emissions, thus producing an autoradiograph. Finally, the autoradiograph was digitized using a desktop scanner.
- *DNA sequencing gel autoradiograph*   Figure 1.9(c) is another type of electrophoretogram, produced at one stage in the DNA-sequencing of gene fragments. In this case, about 50 mixtures were positioned as distinct spots along the top of the gel (as it is currently displayed). Each mixture then migrated down the gel, and DNA fragments produced separate, approximately horizontal bands.
- *Fungal hyphae*   Figure 1.9(d) is a digitized photograph of part of a fungal mycelium *Trichoderma viride*, namely a network of hyphae from a single fungal organism, which was grown on cellophane-coated nutrient agar (Ritz and Crawford, 1990). Image analysis was required here to understand the spatial structure of the fungal hyphae in relation to their environment.
- *Fish*   Figure 1.9(e) is a digitized view of a haddock. The data were obtained by Strachan, Nesvadba and Allen (1990a) as part of their work to automate the identification of fish species for MAFF (Ministry of Agriculture, Fisheries and Food) surveys and for the fish industry. Summary statistics on size and shape need to be extracted from the image, in order to make comparisons between different fish species.

The names we have given some images correspond to their subject matter, whereas in other cases we have named them after the instrument used to gather the data. The choice is somewhat arbitrary, but follows convention, and will be maintained throughout the book.

1.2 WHAT ARE THE DATA?

In this section we shall draw some general conclusions about the examples presented in §1.1.3, in order to answer the question: what are the data?

The array of values of *any* variate measured, either directly or indirectly, at regular points on a two-dimensional grid or lattice may be regarded as a computer image, and analysed using the methods to be presented in this book. Let us first illustrate what we mean by a variate, before considering what the lattice is.

The *variate* may be a measure of the intensity of transmitted light, as in the examples of fungal hyphae, muscle fibres and cashmere fibres. It may be a measure of reflected light, as in the examples of fish and Landsat. It could also depend upon reflected or transmitted radiation in another part of the electromagnetic spectrum (soil aggregate, SAR and X-ray CT), or it could measure emitted radiation after radioactive labelling (electrophoretograms), or reflected ultrasound. It may be measurements of proton density (MRI) or of interference microscopy (algal cells). It could also be measurements of ink uptake and deposition on paper (turbinate bones), or of height, colour, texture, distance etc.

If the object is *three dimensional* then an image may be obtained after physically taking a cross-section (as in the muscle fibres, soil aggregate and turbinate bone examples), or by a computer reconstruction based on some physical property of the object (MRI, X-ray CT and ultrasound). Alternatively, an object could be imaged simply by viewing it from a particular direction. The object could have an opaque surface (fish, Landsat, SAR) or be semi-transparent (algal cells, cashmere fibres). In both cases there may be a focal plane outside which parts of an object are blurred. Some sensors, such as confocal microscopes and magnetic resonance imagers, can collect three-dimensional arrays of data. Such datasets are extremely large and require powerful computers to handle them. They are beyond the scope of this book, but can be analysed using similar methods.

We will now consider what we mean by a *lattice*. Although most variates are measureable at all points within a defined area, a variate can only be recorded at a finite number of them. It simplifies matters if these points are arranged systematically in some form of lattice. Lattices are usually square, although they can be rectangular (as in the ultrasound image) or hexagonal (Serra, 1982) or the points can even be arranged in concentric rings (Silverman *et al.* 1990). Each data value, or picture element (pixel for short), is the average value of the variate for a small region around a point in the lattice. For example, Fig. 1.10(a) shows a region in the bottom centre of the DNA sequencing gel autoradiograph, whereas Fig. 1.10(b) shows a display of the digitized image produced by a desktop scanner. (This is the last we shall see of any source image — from now on, the DNA image and all the other images considered in this book are digital.) Typically, averaging extends further

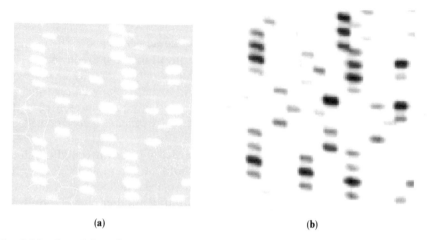

(a)   (b)

Fig. 1.10   Detail from bottom centre of DNA image: (a) original autoradiograph; (b) digitized version.

than the distance between pixels, and images appear slightly blurred when viewed at high resolution. This is the case for all the images in Figs 1.6–1.9. However, for some sensors the extent of averaging is less than the interpixel distance, and details of objects can be lost in the gaps between pixels (Purll, 1985). For example, Glasbey, Horgan and Hitchcock (1994) found this to be the case if their desktop scanner was used to sample photographs at too coarse a resolution.

Although we have said that any variate measured on a lattice *can* be analysed using image analysis techniques, not every variate should be. Plots in a field experiment are often arranged spatially in a two-dimensional grid, so the crop yields of the plots could be regarded as an image. However, it is far more sensible to use statistical methods such as analysis of variance to interpret these data. Also, variates that vary smoothly over space, barometric pressure for example, do not lend themselves to image analysis. All the examples in Figs 1.6–1.9 show abrupt changes in pixel value, which are interpreted by our eyes as boundaries or as the edges of objects. These are characteristics that variates will tend to have if they are to be successfully analysed using the methods in this book.

Terminology

Throughout the book we will consider an image to be a two-dimensional array of numbers and will denote pixel location by (i, j), where i is the *row index*, an integer ranging from 1 to n, and j is the *column index*, similarly between 1 and n. The number of rows may be different from the number of columns, resulting in images that are rectangular rather than square (in fact,

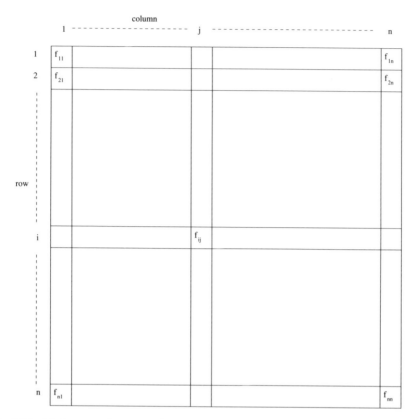

Fig. 1.11   Mathematical notation for digital image.

some of our examples are) but to keep the mathematics as simple as possible we shall use a single dimension, n. The reader should find it relatively simple to generalize to rectangular images. We shall call the value of the variate (or variates) at a pixel the *pixel value* and denote it by f_{ij}. By convention, indices begin in the top-left corner of images so that, for example, in the digitized image of the Mona Lisa, $f_{11} = 132$, $f_{12} = 128$ and $f_{21} = 139$ (see Table 1.1). We shall regard pixels as *discrete points* in the imaging plane, although we shall display them as *square blocks*. Figure 1.11 summarizes the notation.

Pixel values are *univariate* if only one measurement is made at each pixel, otherwise they are *multivariate*. If univariate then they may simply take values of 0 and 1, in which case the image is *binary* (the turbinate image, for example), or they may be multilevel, termed *greyscale* (for example, all the individual images in Figs 1.6–1.9). Because images typically contain a large number of pixels, it is convenient to keep the storage space as small as possible for each pixel, so values are often rounded to integers in the range 0 to 255—which occupy one byte in a computer. (This is the case in Table 1.1.)

Table 1.2   Data sets and displayed ranges of values in Figs 1.6–1.9.

Image[a]	Number of		Pixel range		Displayed range	
	rows	columns	min.	max.	black	white
Algal cells	512	512	23	236	23	236
Cashmere fibres	256	256	35	219	35	219
DNA sequencing gel	512	300	29	226	29	226
Electrophoretogram 1	370	400	7	249	7	249
Electrophoretogram 2	370	400	10	213	10	213
Fish	200	500	23	255	23	255
Fungal hyphae	500	500	11	244	11	244
Landsat band 1	512	512	61	254	61	100
Landsat band 2	512	512	21	134	21	55
Landsat band 3	512	512	17	165	17	60
Landsat band 4	512	512	8	198	8	140
Landsat band 5	512	512	2	254	2	105
Landsat band 7	512	512	1	215	1	60
MRI inversion recovery	128	128	0	253	0	130
MRI proton density	128	128	0	253	0	253
Muscle fibres	512	512	7	218	7	218
SAR[b]	250	250	0.3	10445	0.3	100
Soil aggregate	512	512	3	164	3	80
Turbinate bones	530	460	0	1	0	1
Ultrasound	300	360	8	239	8	239
X-ray CT[b]	256	256	−1000	834	−1000	834

[a] All images, except for Landsat, are available by anonymous FTP from Internet site peipa.essex.ac.uk. In case of difficulties, contact the publisher.
[b] In uses of these images after Chapter 2, the SAR data are log-transformed and the X-ray CT data are restricted to the pixel range −250 to 260.

The sizes of the images in Figs 1.6–1.9, and details of pixel and display ranges, are summarized in Table 1.2. In most instances, we displayed the smallest pixel value as black and the largest value as white, with shades of grey used to represent intermediate pixel values. However, where this produced a display that was too dark for much detail to be discerned, the largest 1% of pixel values were all displayed as white and a contrast stretch was used to increase the brightness of all other pixels (see §2.2.2).

Multivariate images can be categorized into several types:

• *Multispectral*: the variates measure intensity in different parts of the electromagnetic spectrum (as in the Landsat example and in the use of red, green and blue light in colour images).

• *Multimodal*: the variates measure different physical properties of the same object (MRI inversion recovery and proton density, for example).

• *Multitemporal*: a single variate is measured at different times, possibly seconds or years apart.

It is important in any practical application of image analysis to consider whether, by collecting data in a different way, image interpretation could be made simpler. For example, variations in illumination can sometimes be dealt with by collecting two images, one with the specimen present and one of the background alone, then forming a pixel-by-pixel ratio. In microscopy, samples could be stained or viewed using a different microscope modality or different lighting conditions. Strachan *et al.* (1990a) had an elegant solution to the problem of distinguishing between the fish and its background in Fig. 1.9(e), a task that looks simple to the human eye but is difficult to program a computer to achieve. They collected two images: one was obtained with illumination from above (Fig. 1.9e), and the other with backlighting only, which showed the fish as a dark object on a white background. By combining the two images, they were able both to distinguish between the fish and its background and to record surface details of the fish.

In this book, we assume that the reader is mainly interested in analysing images of static objects in laboratory-type situations where analysis time is not critical, samples can be prepared to some extent and processing can be semi-automatic. Real-time, fully automatic image processing of biological objects is also of importance, for example in grading of agricultural produce, but is beyond the scope of this book. Topics in computer vision, such as motion, 3D modelling, real-time processing and high-level image interpretation in the domain of artificial intelligence, will not be dealt with. Nor shall we consider the mathematics of image reconstruction from projections, such as that used to produce the X-ray CT image. The reader is referred to texts such as Rosenfeld and Kak (1982), Gonzalez and Wintz (1987) and Jain (1989) for coverage of these topics.

In the following section, we give an overview of how computers can be used to analyse the types of images we are considering.

1.3 WHAT DOES IMAGE ANALYSIS CONSIST OF?

One way of describing image analysis is to recognize five distinct stages that follow each other logically, namely display, filters, segmentation, mathematical morphology and measurement. To illustrate the stages, consider the algal image (Fig. 1.6a).

• *Display* of an array of pixel values as a picture on a computer screen or on paper, as we have already seen in this chapter, is the first stage in analysing a digital image. Magnification of pixels is also useful for seeing details in an image. Figure 1.12(a) shows two cells from the algal image in close-up.

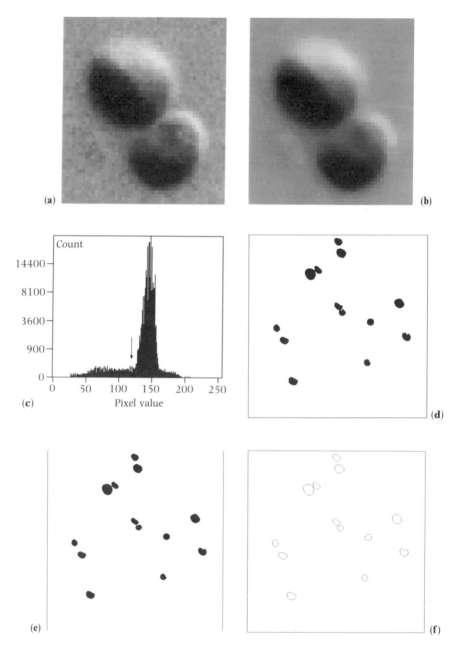

Fig. 1.12   Stages in image analysis illustrated using algal image: (a) detail from image; (b) same detail after application of 5 × 5 moving median filter; (c) histogram, on a square-root scale, of pixel values after filtering, with an arrow to indicate the threshold; (d) result of thresholding image at pixel value 120, to produce a binary image; (e) result of applying morphological opening to (d); (f) separated objects in (e) counted.

- *Filters* enhance images by applying transformations based on groups of pixels. They are computationally efficient methods for reducing 'noise' levels in images and emphasizing edges. (*Noise* is a technical term used by statisticians and engineers, among others, to describe disturbances in data that are either uninterpretable or not of interest.) The moving-median filter smooths flat regions of images, but preserves edges, by replacing each pixel value by the median of the values in a specified local region. Figure 1.12(b) shows the result of applying a 5 × 5 median filter to the same section of the algal image as Fig. 1.12(a). By comparing Figs 1.12(a) and (b), we can see that the noise that caused the speckle has been reduced.

- *Segmentation* divides an image up into regions, which correspond to different objects or parts of objects, by classifying all pixels. In the algal image, the aim is to identify which aggregations of pixels belong to which cells. Figure 1.12(c) shows the distribution of pixel values in the median-filtered image, displayed on a square-root scale. Background pixels (those not belonging to cells) contribute the bulk of this distribution, with values in the range 130 to 160. By thresholding at 120, that is labelling all pixels with values less than 120 as belonging to cells, dark parts of the algal cells are distinguished from the background, as shown in Fig. 1.12(d).

- *Mathematical morphology* uses a collection of operations to study the shapes of objects, as expounded mainly by Serra (1982). For example, two pairs of cells are touching in Fig. 1.12(d), and boundaries are slightly roughened due to noise, even though the algal cells themselves should be smooth. Figure 1.12(e) shows the result of applying a morphological operation called an opening. Now all the cells are separated and have smooth boundaries.

- *Measurement*, the extraction of quantitative information from images, is the final stage in the analysis. The number of separate objects can be counted, as shown in Fig. 1.12(f). Although the algal cells are approximately circular in shape, these objects are more ellipsoidal because only the darker sides of the cells have been identified. By measuring the maximum distances between pixels in each object, the cell diameter can be estimated. Cell 4 appears to be largest, with a diameter of 33.6 pixel units, while 12 appears smallest, diameter 19.0.

Analysis of any particular image is likely to require several of these stages, in this order, but sometimes re-using techniques from previous stages. For example, after filtering, it is usual to display the transformed version of the image. Of course, the way in which data were collected (§1.2), and the questions to be answered, are of crucial importance in determining how a particular image should be analysed.

There are other ways of approaching image analysis. One alternative is where data are processed in a single step to produce the final result. This is the approach taken by Grenander, Chow and Keenan (1991) in fitting template models to images of hands.

In the five chapters that follow, we shall deal in greater detail with the five stages identified above. Each chapter is reasonably self-contained, so the reader need not work through the whole book in order. Finally, in Chapter 7 we shall review the progress we have made in analysing each of the 14 datasets, and give pointers to further work.

1.4 SUMMARY

The key points of this chapter are as follows:

- Image analysis is simply the extraction of information from pictures.
- The human vision system is superb, but qualitative.
- Computers can do better than humans at extracting quantitative information, and can reduce the tedious aspects of image interpretation.
- The images to be considered in this book are drawn from microscopy, medical scanning systems, direct photography and remote sensing.
- Any variate measured at regular points on a two-dimensional grid may, after digitizing, be regarded as a computer image.
- The pixel value at location (i, j) will be denoted by f_{ij}. For simplicity, images will be assumed to be square, with both the row index i and the column index j ranging between 1 and n.
- Images may be univariate or multivariate, and each variate may be binary or greyscale.
- Image analysis can be subdivided into five stages: display, filters, segmentation, mathematical morphology and measurement. These are the topics of the following five chapters.

2

Display

It is normal in any area of science that when a new specimen is acquired, it is examined carefully before any measurements are taken. The same approach is applied to digital images. If many similar images are to be dealt with, we may not wish to spend time looking at all of them. We may in fact be wanting to use digital image analysis in order to avoid this. But we shall still wish to spend some time looking at images while developing a method to deal with them in large numbers.

One might ask why we don't simply look at the object that was the source of the digital image. There are a number of reasons for this:

- It may not be possible. The original object may not be available for examination, and the digital image may be all we have. For example, satellite imagery exists in digital form only, and the same is true of many types of non-invasive medical imaging. If we want to look at the object, we must look at the digital image.
- Even if the original object is available for viewing, we ought still to look at the digital version with which we shall be working. This will allow us to see what level of detail is retained, how much blurring is present, whether there are problems in the image capture mechanism, what difficulties there are likely to be in producing an automatic analysis of the image, and so on.
- It may be possible by image enhancement to see aspects of the image that are difficult to spot in its original form.

This chapter considers the issues involved in producing a display from the various types of image that we may be working with. The display of binary, greyscale and multivariate images are considered in §§2.1, 2.2 and 2.3 respectively. This chapter will also consider enlarging and reducing an image for convenience of examination (§2.4) and manipulating two or more images of the same scene or object so that features are at the same pixel position in all of them (§2.5). The main points of the chapter are summarized in §2.6. Since the display will use some form of computer equipment, we shall need to make some references to possible limitations in computer hardware. (For example, one obviously cannot produce any form of colour display on a monochrome monitor!) A consideration of what is wanted from the display component of an image analysis system should be considered when choosing one. Printing an image on paper should also be considered as a form of display. General guidelines on matters of hardware and software are to be found in the Appendix.

2.1   BINARY DISPLAY

A binary image is the most straightforward type to display. Pixels take only values of 0 and 1. The display simply needs to distinguish between these two levels. A natural way to do this is to show one value as black and the other as white.

Any computer monitor capable of displaying images can do this. If the digital image is greyscale, rather than binary, a threshold must be chosen if a binary display is required: pixel values on one side of the threshold are displayed as black, those on the other side as white. Figure 2.1 shows a thresholded display of the fungal image. (Choosing a threshold is addressed in §4.1.) We see that almost all the detail in the original (Fig. 1.9d) is retained in this binary display.

Fig. 2.1   Binary display of thresholded fungal image, using a threshold of 150.

Which pixel value should be shown as black and which white? Sometimes one choice will seem more natural. After thresholding a greyscale image, it is natural to show the pixel values that are below the threshold as black, and those above as white. However, one easily adapts to a form of display that at first seems unnatural. Astronomers generally look at negative photographic images of the night sky, in which stars are black on a white background.

2.2. GREYSCALE DISPLAY

Most common digital images—certainly most in this book—are greyscale. The natural way to display such images is to use the pixel values to specify the brightness with which a pixel is illuminated on a computer screen, or how bright the pixel appears on a printed page. As described in §1.2, the pixel values are some measured physical property of the object being studied. If this property is the amount of reflected or transmitted light then the display we produce will simply look like a black and white photograph of the object. We shall refer to such a display as greyscale, sometimes termed monochrome, rather than the more colloquial 'black and white', to distinguish it from binary images where each pixel is purely black or white.

Often, the pixel values are some other physical property of the image, and showing larger (or, if preferred, smaller) pixel values as brighter is simply a device to enable us to see the spatial structure in what has been measured. This is not new in digital image analysis. It is standard practice, for example, to use brightness in displaying X-ray images, where brightness indicates how opaque a region of the object is to X-rays.

In using brightness in an image display, we have control over what level of brightness we assign to each pixel value. We can vary the contrast and overall brightness in ways that help us to see what is of interest in the image.

2.2.1 Perception of brightness

We have referred to a multilevel univariate image as a greyscale image because we use levels of greyness to display its values. The term 'grey' implies in effect that there is no colour present: all parts of the visible spectrum are equally represented. (Colour is discussed in §2.3.) A physical measurement of the amount of light is its intensity. This is proportional to the energy in the light, and is the square of the amplitude of the light waves.

The human eye is not simply an instrument that registers the amount of light it receives. We tend to see zero intensity as black, low intensities as dark grey, with the greyness lightening as the intensities increase. The lightest objects we can see at any time tend to be perceived as white. However, the eye is very sensitive to the immediate surroundings of an object. This is best demonstrated by the familiar optical illusion shown in Fig. 1.4(b). The main lesson to be derived from this is that the eye is not to be trusted for objective assessment of the absolute intensity of different parts of an image display.

Generally, there are a finite number of grey levels to which each pixel on a computer monitor can be set. On old monitors, just two levels were possible, meaning that only binary images could be displayed. On most modern monitors, at least 256 grey levels are possible. (Note that monitors are themselves analogue rather than digital, but appear digital because of the digital electronics that drive them.) If the number of display levels on a monitor is less than the number of distinct pixel values in the image then some detail will be lost, since different pixel values will have to be shown as the same brightness on the monitor. However, 256 grey levels should be adequate for greyscale images.

It should not be assumed that the intensity response of a monitor is linear in the nominal values to which it is set. This is something which can be assessed by comparing, for example, an image in which all pixel values are half the maximum possible value with one constructed as a fine chessboard pattern of maximum values and zero. These two patterns should emit the same amount of light on average, and should be perceived similarly from a distance. Often, they are not.

2.2.2   Display enhancement

The most straightforward way to display a digital image is by a linear mapping of the range of pixel values onto the range of brightness intensities to which the monitor can be set. Sometimes the mapping is trivially simple. Often, both the pixel values and the possible display intensities will be integers in the range 0 to 255. If so, we simply display each pixel value as that display intensity.

A one-to-one display of the pixel values in an image may not be the best for seeing what is of interest. We may be able to see more by transforming the intensities displayed. This is analogous to the variable transformation (such as taking logs) that we might use with any other type of scientific data. A transformation would be appropriate if, for example, most of the pixel values were small, with a few large values. The differences between large and small values would swamp any subtle differences between small values. If the variation in the small values is of interest then a transformation so that these small values occupied a wider range of display intensities would be of help.

A transformation may be effected in two distinct ways. One is to transform the pixel values stored in the computer. The other is to manipulate the way in which particular pixel values are displayed. An important concept in many image display systems is that of a *look-up table* (LUT). This is a listing of the intensity on the monitor to be associated with each possible pixel value. Altering the LUT will have the same effect on the display as if the pixel values themselves were changed. It has the advantages of being much quicker in computer

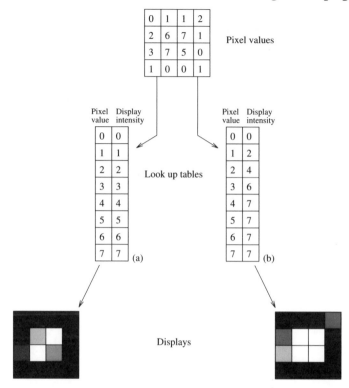

Fig. 2.2   Use of look-up tables for displaying pixel values. Both pixel values and display intensities are in the range 0–7. In (a) the LUT produces display intensities proportional to the pixel values, whereas in (b) they are modified to emphasize variation in small pixel values.

time (since every pixel in the image need not be transformed) and none of the detail in the image is lost, this being a possibility when some range of values is compressed. Of course, the possibility of transforming the original pixel values before subsequent analysis still remains. Use of LUTs is illustrated in Fig. 2.2.

In the rest of this section, we shall assume, without loss of generality, that the LUT is being changed when a transformation is made. A number of transformations are commonly used.

Piecewise linear

The *piecewise linear* transformation function consists of linear segments. It is sometimes referred to as a *contrast stretch*. Often the linear segments will be of the special form illustrated in Fig. 2.3. A selected range of pixel values is allocated the complete range of display intensities. Here they are applied to

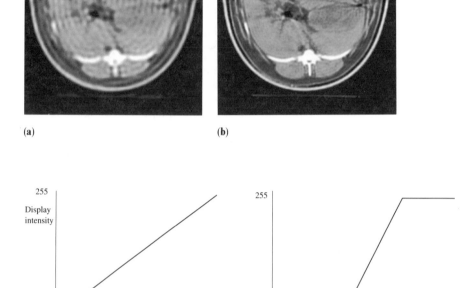

(a)   (b)

(c)   (d)

Fig. 2.3   Piecewise linear intensity transform: (a) original X-ray image; (b) display chosen to emphasize detail in part of the range of image intensities; (c) and (d) the LUTs (as plots rather than lists of intensities) used for (a) and (b).

the X-ray image. The original image has pixel values between -1000 and 834. Fat tissue has values about -100 and muscle tissue about 30, so these do not differ much in the original display. If the values between -250 and 260 are stretched to cover the whole range of display intensities, much greater detail can be seen in the internal tissues of the sheep. We shall use this transformed version of the image in preference to the original.

The transformation has the effect that pixel values between two limits, say a and b, are allocated the full range of intensities available on the monitor. Anything above or below the range a to b is shown as full or zero intensity respectively. Assume 0 and I_{max} are the minimum and maximum intensities available on the display. If f_{ij} is the pixel value at location (i, j), the display intensity I_{ij} is given by

$$I_{ij} = \begin{cases} 0 & \text{if} \quad f_{ij} \leq a, \\ I_{\max} \dfrac{f_{ij} - a}{b - a} & \text{if} \quad a < f_{ij} < b, \\ I_{\max} & \text{if} \quad f_{ij} \geq b. \end{cases}$$

Values for a and b may be chosen in a number of ways. These are illustrated in Fig. 2.4 for band 2 of the Landsat image.

- We may set a to be the minimum of f over the image, and b to be the maximum. This ensures that detail in the extremes of the range of pixel values is

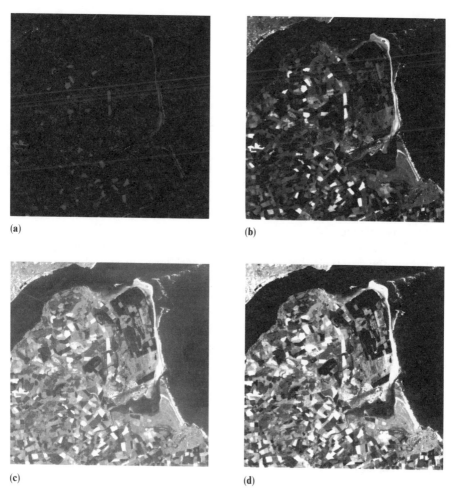

(a)

(b)

(c)

(d)

Fig. 2.4 Piecewise linear contrast stretches applied to band 2 of the Landsat image. The values of a and b are chosen (a) as the minimum and maximum pixel values in the image, (b) as $\bar{f} \pm 2\frac{1}{2}\sigma$, (c) as the 5th and 95th percentiles, and (d) manually.

not lost. This is illustrated in Fig. 2.4(a). The minimum pixel value is 21, and a few very large pixel values in the urban area near the top left give $b = 134$ and as a result most of the image remains dark.

- If we are prepared to lose some detail in the extremes, we could set

$$a = \bar{f} - c\sigma,$$
$$b = \bar{f} + c\sigma,$$

where $\bar{f}$ is the mean pixel value in the image, σ is the standard deviation of the pixel values, i.e.

$$\sigma = \sqrt{\frac{1}{n^2} \sum_{i=1}^{n} \sum_{j=1}^{n} (f_{ij} - \bar{f})^2},$$

and c is a multiplier such as 2 or $2\frac{1}{2}$. This will be useful if there are a few extreme pixel values, which would otherwise force the display intensities for the majority of pixels into a small range. This is illustrated in Fig. 2.4(b), where we get $a = 15$ and $b = 45$, using a multiplier of $2\frac{1}{2}$. This transform shows improved contrast over most of the image.

- A closely related choice is to set a and b to the 5th and 95th percentiles (for example) of the distribution of pixel values. (The pth percentile is defined to be the number F for which $p\%$ of pixel values f_{ij} are less than F.) This is illustrated in Fig. 2.4(c). Here $a = 24$ and $b = 41$, and even stronger contrast can be seen than in Fig. 2.4(b), where $\bar{f} - 2\frac{1}{2}\sigma$ extended beyond the range of pixel values.

- a and b can be chosen manually. This was done in Fig. 2.4(d). Here $a = 25$ was chosen as a typical pixel value in the dark forest area in the top right of the image, and $b = 56$ as a typical value in the bright fields of oil-seed rape scattered throughout the image. This contrast stretch makes it apparent just how different these fields are from the rest of the landscape, a fact that was lost in Fig. 2.4(c), since they represent less than 5% of the image.

Another type of piecewise linear transform is one that consists in simultaneously stretching several small ranges of pixel values to the full intensity range, so that a plot of the LUT looks like a series of steep ramps. This is useful for revealing small local variations in pixel values. Bright and dark areas in the transformed image will follow contours of pixel values.

Exponential, logarithmic

The *exponential* and *logarithmic* functions are examples of transformations that enhance the display of the upper and lower parts respectively of the range of pixel values (see Fig. 2.5). The effect of the logarithmic transform on the SAR

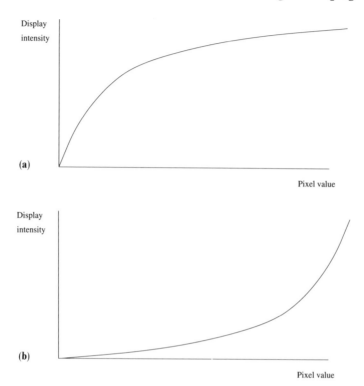

Fig. 2.5   (a) Logarithmic and (b) exponential intensity transforms.

image is shown in Fig. 2.6. In the original image, there was greater speckle variability in the brighter parts of the image than in the darker. After the log transform, the speckle variability should be equal in all parts of the image. This can be shown mathematically from the known properties of radar speckle (Skolnik, 1981; Oliver, 1991). A piecewise linear transformation using the minimum and maximum pixel value was applied after the log transform. The transformed pixel values have been used in subsequent analyses in this book.

Histogram equalization

Histogram equalization is a transformation that ensures that all display intensities are approximately equally represented. The intention is that ranges of pixel values are allocated portions of the display intensity range according to the frequency with which they occur in the image. Figures 2.7(a) and (b) show the cumulative histograms of the intensities of the muscle fibres image before and after equalization, and (c) and (d) show the original and equalized images. We can see more detail in the cells in the equalized image. This is because the pixel

Fig. 2.6   Logarithmic transform of the SAR image.

values in the dark cells, for example, have been allocated a greater range of display intensities than in the original display. The brighter cells also have a greater range of intensities.

To see how a histogram equalization is achieved, let I denote the display intensity, where I can take the integer values $0, 1, \ldots, I_{\max}$, and let $p(f)$ be the proportion of the pixel values $\leq f$. If pixel values f are displayed with intensity

$$I = I_{\max}\, p(f)$$

then the proportion of pixels with display intensity $\leq I$ will be $I/I_{\max}$, leading to a linear cumulative distribution of intensities, i.e. a uniform distribution. In practice, the discrete (i.e. non-continuous) nature of the image histogram will mean that the transformed image will only have an approximate uniform distribution. For a more detailed discussion of histogram equalization, see Gonzalez and Wintz (1987, Ch.4).

2.3   COLOUR DISPLAY

Colour is an important aspect of human vision. It is so fundamental that it is impossible to describe its sensation in terms of anything more basic. We see the

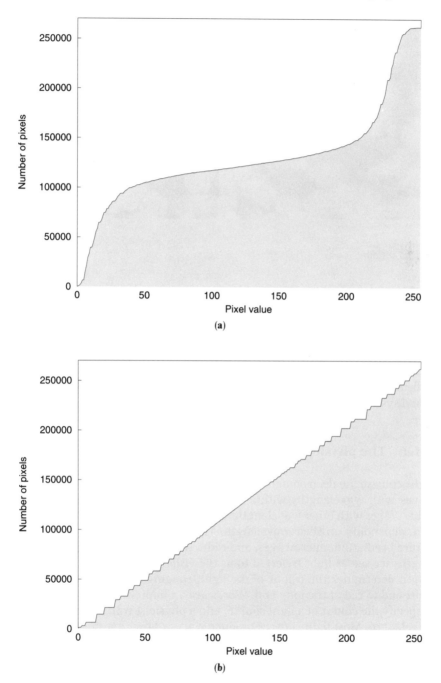

Fig. 2.7 Cumulative histogram of the muscle fibres image (a) before and (b) after equalization. Muscle fibre image (c) before and (d) after equalization.

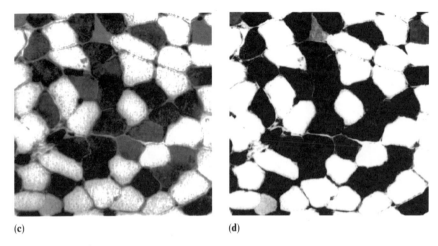

(c)   (d)

Fig. 2.7   *Contd.*

world in colour, and so colour is an aspect of many of the things that we look at in science. Some types of imaging systems using non-light-based physical processes generate images without colour. Colour can be assigned to such images to help us see variability in the quantity measured. It is also a natural way to deal with images in which more than one variable has been measured at each pixel. To understand how colour can be used in display, it is necessary first to understand what colour is physically, and how it is perceived by the human visual system.

2.3.1   The physics and biology of colour

Light consists of electromagnetic waves. The human eye is sensitive only to waves with wavelengths within a certain range, which are termed visible light. Waves with other wavelengths are considered as different types of radiation, depending on their wavelength. These include X-rays, ultraviolet and infrared radiation, microwaves, and radio waves. Within the range of wavelengths we see as light (referred to as the *visible spectrum*), the actual wavelength determines the colour of the light, ranging from indigo/violet at the short end to red at the long end. Words are an imprecise and subjective way to specify the colour of a light source, and a physicist would simply quote the wavelength. Most light sources produce waves not with a single wavelength but with a distribution of different wavelengths.

  The biology of how our eyes see colour is different from the physical reality. It is this biology with which we are most concerned, since we are discussing the use of colour to display images for the human interpreter. The retina of the eye contains cells (referred to as cones) that differ in their sensitivity to light of particular wavelengths. There are three types of cone: one type has its maximum sensitivity to light in the blue part of the spectrum, another type to the green part and a final type to the red part. Light consisting of a single wavelength in the red part of the spectrum will be detected most strongly by the red-sensitive cells, and we see it as red. Light of a single wavelength lying between the green- and red parts of the spectrum produces a response in both the green- and red-sensitive cells, and a sensation of yellow results. A similar sensation is produced by light consisting of a mixture of one red and one green wavelength—it looks yellow—since it produces the same effect on the colour-sensitive cones. Any combination of wavelengths produces a different colour sensation, according to how much of a response it produces in each of the three types of colour-sensitive cells. An equal response in each of the three basic colours, blue, green and red, produces a perception of grey.

Because this is the way colour is perceived, we can regard colour as a three-dimensional quantity. To reproduce any colour, we specify the intensity of red light, green light and blue light, the so-called primary colours. This is referred to as the *RGB* (red, green, blue) system. We can think of the RGB components as specifying a point inside a cube, the sides of which correspond to the range of possible intensities of blue, green and red that can be specified. Most technological use of colour, from photography to television screens and computer monitors, employs an RGB system to produce different colours. For example, a colour monitor has a screen that consists of many small, light-emitting dots of three types—blue, green and red. The brightness of each of the three types of dots determines what colour is seen in a particular part of the screen.

2.3.2   Pseudocolour

One use of colour in image display is to assign different colours to different pixel values in a greyscale image. This is termed *pseudocolour*, since it does not reflect actual colour variation in the scene imaged. Such a display sometimes has the advantage over a standard greyscale display that the eye is better able to see colour differences, and to compare colours in different parts of the display, than to see differences in grey levels.

There are many possibilities for assigning colours to pixel values. We can regard the sequence of colours we assign to the range of pixel values, from its minimum to its maximum, as defining a path in the RGB colour cube. Two possibilities for such paths are shown in Fig. 2.8 (colour plate). Figure 2.9 (colour plate) shows the fish image (Fig. 1.9e) displayed in pseudocolour. Many image features and comparisons can be more readily made in colour than in grey levels, since the problem of sensitivity to the surrounding area (as in Fig. 1.4b) is much reduced. However, pseudocolour is less suitable for images that we are accustomed to seeing in greyscale or natural colour.

A greyscale image of a human face looks unnatural when displayed in pseudo-colour, and is more difficult to recognize.

2.3.3   Direct allocation

If we have two or more variables measured at each pixel in an image (i.e. a multivariate image), we may use them to control different RGB colour components in the display. This is particularly convenient if there are three variables. In this case, we simply assign each of them to one of the three RGB components. If the three variables are in fact measurements of the light in the red, green and blue components emitted or reflected by a real scene then a natural looking colour picture of the scene is produced. This is illustrated in Fig. 2.10 (colour plate) for the satellite image, where bands 1, 2 and 3 are the blue, green and red parts of the visible light spectrum respectively. The three bands have been individually stretched so that $\bar{f} \pm 2\frac{1}{2}\sigma$ occupies the full intensity range as described in §2.2. If the variables assigned to the RGB components do not represent blue, green and red light then the display will not look like anything natural, but the eye will still be able to study its features.

If we have two image variables, we could assign them to two of the colour components and set the third to a constant value. Another possibility is to let two of the components be determined by one of the variables, and the third by the other variable. This is shown in Fig. 2.11 (colour plate) for the two MRI variables. If we have more than three image variables, there are a number of possibilities:

- We may feel that a selection of just three of them will produce a display in which most features of interest can be seen. Some experimentation may be needed to find this combination.
- We could create some derived functions of the variables, and use these to determine the colour components. One possibility here is to use the *principal components* (Krzanowski, 1988, p.53). These are a sequence of linear combinations of the variables which contain the greatest amount of variability while being independent of all previous components in the sequence. We may then produce a display from the first three, or some other set of three components. This does not guarantee a display which has the most image features of interest. There is no certain way to find appropriate combinations of image variables for this purpose.
- The *maximum noise fraction* (MNF) technique (Green *et al.*, 1988) finds a sequence of linear combinations of variables, with the noise-to-signal ratio decreasing through the sequence. Noise is assumed to be lowest when the correlation in pixel value between neighbouring pixels is highest. Later (i.e. lower noise) terms in the sequence can be used to determine colour components.

- Some derived functions of the variables which are thought to reflect important features of the scene may be used. For example, in remote sensing, vegetation indices (Jensen, 1986, Ch.7) are considered to indicate amount or type of vegetation. When based on ratios of variables (which correspond to different parts of the electromagnetic spectrum), they can eliminate or reduce the effect of variations in illumination caused by topography.
- The statistical technique of projection pursuit (Jones and Sibson, 1987; Nason and Sibson, 1991) has been developed to attempt to find interesting projections of multidimensional data.

It can be seen that a great many colours can be obtained on a display unit from the combinations of three basic colour components. If each component can be displayed in 256 different intensities for example (which is often the case) then there are $256^3 = 16\ 777\ 216$ different possible colours that a pixel can have. However, the memory configuration of many computers is such that not all of these can be available at the same time. Many operate by allowing the display intensities to be one of a smaller number (often 256) of different colours. Which colour is assigned to each of the 256 possibilities is determined by a LUT. We must then choose which 256 colours out of the 16 777 216 possibilities to use. A straightforward way to do this is to select eight colours from their full range for two of the RGB components, and four from the other, making $8 \times 8 \times 4 = 256$ colour combinations. This is easily programmed, since 4, 8 and 256 are powers of two. Other possibilities, sometimes implemented in software packages, are to select from the 16 777 216 possibilities according to the observed colour combinations in the image, to minimize the loss of detail when reducing to 256 colours (Heckbert, 1982).

Other colour coordinates

Colour can be specified using other coordinate systems in the colour cube. The most common alternative to the RGB coordinates is the *intensity, hue, saturation system (IHS)*:

- *Intensity* or *lightness* is how bright the light is overall. It may be defined as the projection of a colour in RGB coordinates onto the diagonal from black to white.
- *Hue* indicates angular position of the colour in the plane perpendicular to the black-to-white diagonal. The choice of zero angle is arbitrary.
- *Saturation* measures how far from the diagonal line the colour is, as a proportion of the distance from the diagonal through the colour to the edge of the cube.

This is illustrated in Fig. 2.12. If we assign three of the variables in a multivariate image to defining intensity, hue and saturation, the above definition allows us to obtain the RGB components needed for the monitor.

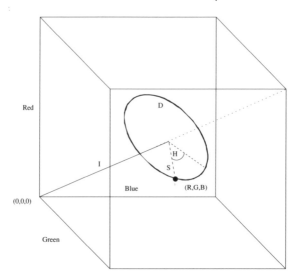

Fig. 2.12   Illustration of the intensity, hue and saturation coordinates for describing colour. The point (R, G, B) lies on the disc D, which is perpendicular to a diagonal line drawn from black (0,0,0) to white (1,1,1). The intensity I is the distance from (0,0,0) to D. Hue H is the angle in D that the point (R, G, B) makes with a line drawn towards blue (0,0,1). Saturation S is the distance of the point (R, G, B) from the centre of D as a proportion of the distance along a line from the centre of D through (R, G, B) to the edge of the cube.

Alternatively, we use the definition to describe any colour specified by RGB coordinates in terms of IHS coordinates, which may have more intuitive appeal. Algorithms for performing these coordinate transformations may be found in Foley *et al.* (1991).

   The IHS coordinates can be used in other ways than simply assigning three components of a multivariate image to intensity, hue and saturation. If we have two variables, we might assign one to intensity, one to hue and set saturation to its maximum. With one variable, we might assign it to hue, set saturation to its maximum and intensity to some value between zero and its maximum. This leads to a form of pseudocolour (as shown in Figs 2.8b, d). This is illustrated in Fig. 3.18 (see §3.4.2). Note that we cannot use maximum intensity, since this leads to an all-white image regardless of hue and saturation. Hue is undefined when red, green and blue are equal.

   Other colour coordinate systems also exist. Some of these have been defined to take account of the properties of monitors or of human perception of colour. A description of some of these may be found in Foley *et al.* (1991). If an image is being captured in colour, for example with a colour digital camera, it can be important to understand its colour response and to calibrate it with respect to some colour coordinate system. For an example, see Strachan, Nesvadba and Allen (1990b).

A final point to note here is that there are some differences between how images are displayed on monitors, and how they are produced on paper. A great many different devices for printing images exist, with varying quality of reproduction. The quality is often poorer than the display on a monitor.

2.4 ZOOMING AND REDUCTION

When an image has fine detail, we may not be able to examine this detail easily in a standard display on a monitor. We should like to enlarge a part of the image. This is known as zooming. We cannot usually enlarge the pixels on the monitor, and zooming must consist of allocating more than one pixel on the monitor for each pixel in the image.

The simplest form of zooming is *pixel replication*. For example, if we display an image pixel using a block of $k \times k$ pixels on the display, then we will apparently have increased the size of the image (see Fig. 2.13). Clearly, no more detail is present, although it may be easier to see what is there. Of course, there may no longer be room on the display for the whole image, as in

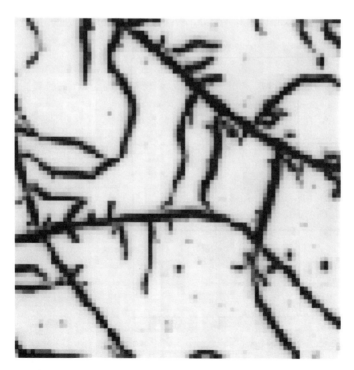

Fig. 2.13   Zooming to enlarge part of the fungal image (Fig. 1.9d). An enlargement of the top right of the original image is shown. Each pixel is repeated in a 5 × 5 block.

Fig. 2.13. It is equivalent to holding a magnifying glass in front of the screen. It is not always necessary to replicate pixel values in the computer memory. Some computer hardware has the ability to replicate pixels on the screen by making use of the way that the display control electronics access the computer memory. Pixel replication is satisfactory if we wish to magnify the image by integer ratios. If we wish to magnify by fractional ratios, there is the possibility of replicating pixels by different amounts in some pattern. For example, if we replicate every second pixel only, and every second row, we will have magnified the image by 1.5. This gives an acceptable result if we are concerned only with the larger features of the image, but distortion of small-scale details can be noticeable.

On other occasions, we may wish to *reduce* the size of an image. This may be for display purposes, because our original image has too many pixels to be accommodated on the monitor. Alternatively, we may wish to reduce the size of an image so that it can be processed in less computer time, or occupy less storage space. (Image compression is also considered in §A.1.4.) Reducing the image we intend to work with will involve some loss of detail. We may reduce the size of an image by integer ratios in two straightforward ways.

• We can form the reduced image by selecting a subset of pixels from the original image. For example, if we take every third pixel in every third row of the

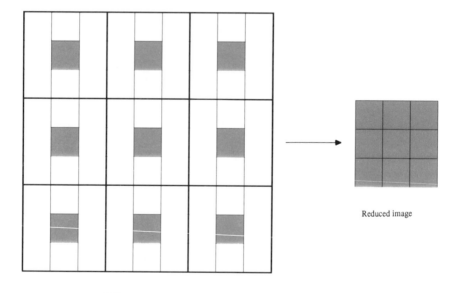

Reduced image

Full image

Fig. 2.14 Image reduction. In reduction by pixel sampling (shown on right), the shaded squares in the full image are used to form the reduced image. In reduction by block averaging, the average intensity in the blocks shown is used.

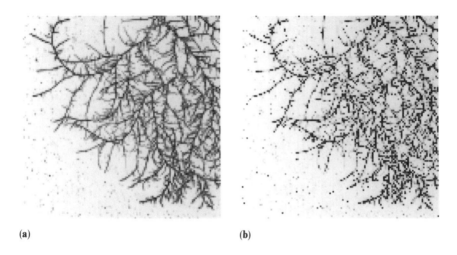

(a)   (b)

Fig. 2.15 Effect of (a) block averaging and (b) pixel sampling in reducing the resolution of the fungal image. Here the size is reduced by four horizontally and vertically.

original image, we shall have reduced the image size by a factor of nine. This is known as *pixel sampling*.

- We may define each pixel in the reduced image to be the average of a three-by-three block of pixels in the original image. This reduces the loss of information, unlike the equivalent pixel sampling, which throws away 8/9 of the pixels in the original image. This is known as *block averaging*.

The difference between the two methods is illustrated in Fig. 2.14. The effect of the two methods is demonstrated in Fig. 2.15. If neighbouring pixel values are similar, there will be little difference in these two methods. Pixel sampling has one advantage in that it requires less computer time. For non-integer reductions, sampling and averaging procedures can be devised, with distortion problems similar to those found in zooming.

For both zooming and reduction, particularly with non-integer factors, *interpolation* provides a method that produces smoother results with fewer distortions than simple pixel replication or sampling. The basic idea is that we regard the digital image as a discrete version of some continuous variable, which we have sampled at points on a lattice. To enlarge or reduce the image, we sample at a greater or lesser frequency. The pixel locations in the new image will fall at different positions on the continuous surface. We do not of course know what the pixel values at these positions should be, but we may estimate them by assuming that there is a smooth change in pixel value between the four nearest pixels in the original image. This is illustrated in Fig. 2.16 for the one-dimensional case. If we assume that the inter-pixel distance in the original image is 1, and that we wish to estimate the value f'_y

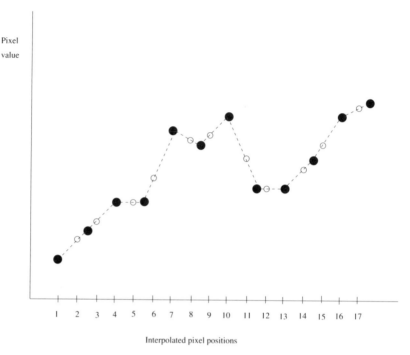

Pixel
value

Interpolated pixel positions

Fig. 2.16   Interpolation in one dimension. If the filled circles denote the original pixel values then linear interpolation will assign the values on the dotted line to the inter-pixel points. If, for example, an enlargement by a factor of 1.5 is required then the open circles are used to supply pixel values. At every third pixel, they coincide with original pixel positions.

at a pixel in the interpolated image, then we have

$$f'_y = (1 - a)f_i + af_{i+1},$$

where i labels the nearest original image pixel $< y$, and $y = i + a$. In two dimensions, we may use bilinear interpolation based on the four nearest pixels in the original image, leading to the formula

$$f'_{yx} = (1 - a)(1 - b)f_{ij} + (1 - a)bf_{i,j+1} + a(1 - b)f_{i+1,j} + abf_{i+1,j+1},$$

where the position (y, x) in the interpolated image corresponds to position $(i + a, j + b)$ in the original image, with $0 \le a, b < 1$.

2.5   REGISTRATION

If we have several images of the same scene, which we wish to analyse jointly, it simplifies matters if they match exactly, i.e. each pixel position corresponds to the same part of the scene in the different images. The images may be different variables obtained with the same instrument, or the same scene viewed with different instruments, under different conditions or at different times. Sometimes the images will not exactly coincide, in which case they must be distorted in order to match. This process is known as *image registration*.

In order to distort images to match each other, we must have an accurate knowledge of the differences between them. This may be expressed in terms of which parts of the digital images refer to which particular parts of the scene, and should therefore be aligned in the two images. Occasionally this may be known *a priori* from the nature of the imaging instruments. More often it will not, and we have to discover what the image differences are. This is usually done by finding *control points*, sometimes called *landmarks*. These are points in the scene whose pixel positions can be obtained precisely in all of the images to be registered. Often they are corners of objects or intersections of narrow straight line or curved features. Enough control points are needed so that the pattern of their positions captures all of the significant differences in object feature positions between the images. Control points may be identified manually, but automatic methods for finding what distortion is needed for images to match have also been proposed—see for example Herbin *et al.* (1989) and Rignot *et al.* (1991)

The next stage is to model the differences between the images. To simplify matters, let us assume that we have just two images. We select one of the images, and model the position of features in the second image as some function of their position in the first image. A pixel with coordinates (i, j) in the first image will correspond to row and column coordinates $u(i, j)$ and $v(i, j)$ in the second image. If we have more than two images to register, we can register all of them to one of the images, which reduces to repeating two-image registrations, or register all of the images to a configuration based on the average position of the control points in the different images.

A simple model for the distortion is the *affine model*, which assumes that the distortion consists only of translation, rotation, size scaling in the horizontal and vertical directions (possibly by different amounts) and shear (which is what turns a square into a diamond shape). It is approximately the distortion you would see if you rotated this book and viewed it from different distances and angles. The distortion may be modelled by

$$u(i, j) = a_0 + a_1 i + a_2 j,$$
$$v(i, j) = b_0 + b_1 i + b_2 j.$$

The coefficients $a_0, a_1, a_2, b_0, b_1, b_2$ may be estimated from the control points by linear regression (Draper and Smith, 1981).

If the distortion is nonlinear then a more complex function is needed. One generalisation is to use a quadratic or higher-order polynomial function.

A quadratic function would have the form

$$u(i,j) = a_0 + a_1 i + a_2 j + a_3 i^2 + a_4 ij + a_5 j^2,$$
$$v(i,j) = b_0 + b_1 i + b_2 j + b_3 i^2 + b_4 ij + b_5 j^2.$$

and this may also be estimated by linear regression. Another possibility is to use thin-plate spline functions (Bookstein, 1989). These are intended to provide an exact fit to the distortion in the control points, subject to minimizing a function (the 'bending energy') of the curvature.

The next stage of image registration is to modify one of the images so that it matches the other. This is often referred to as *unwarping* (and sometimes, confusingly, as *warping*). If we have identified the distortion function as defined above, we unwarp the second image so that it matches the first. To do this, we take each pixel position (i,j) in the first image, and assign to it the image value at position $(u(i,j), v(i,j))$ in the second image. In general, $(u(i,j), v(i,j))$ will not define an integer position in image two, and so will not correspond to a unique pixel value. We have a number of options for how to obtain a pixel value:

- We may take the value at the pixel that is nearest to $(u(i,j), v(i,j))$ by rounding. This is known as *nearest-neighbour* interpolation. It is computationally the quickest, but in some circumstances produces unsatisfactory results. Pixels may be replicated (be assigned to more than one pixel in the unwarped image) by different amounts.
- *Bilinear interpolation* as described in §2.4 uses the four pixels nearest to $(u(i,j), v(i,j))$
- *Cubic convolution* (Bernstein, 1976) requires the fitting of a cubic polynomial to the nine nearest pixels.

A discussion of these issues, and image registration in general may be found in Rosenfeld and Kak (1982, Ch.9).

Figure 2.17 (colour plate) shows the registration of the electrophoresis image of Fig 1.9(a) to that of Fig. 1.9(b). This was based on nine control points at the centre of protein spots which have been found not to vary in different genetic strains of the malarial parasite. The affine transformation and nearest neighbour interpolation was found to give acceptable image registration for these images. Figure 2.17 can be used to find which proteins are the same in both strains, and which are different. This application is discussed in greater detail in Horgan *et al.* (1992).

2.6 SUMMARY

- Images need to be examined before an automatic analysis is produced.
- Different displays are appropriate, depending on whether the image is binary, greyscale or multivariate.

- Binary display is straightforward: we show one pixel type as black, the other as white.
- To display greyscale images, we allocate a display intensity to each possible pixel value. This generates a look-up-table.
- Transforming the pixel values, directly or only in the look-up-table, may help features to be seen. Useful transformations are
  — Piecewise linear (to allocate the full display intensity range to a limited range of pixel values).
  — Exponential or logarithmic (to emphasize variation in large or small pixel values respectively).
  — Histogram equalization (to allocate ranges of display intensity according to frequency of occurrence of pixel values in the image).
- Multivariate images are best displayed in colour, which can be based on the human red, green, blue (RGB) system of perceiving colour.
- Pseudocolour can help in examining a greyscale image.
- Zooming may be done by pixel replication, if we wish to enlarge by an integer ratio, or by interpolation otherwise.
- Reduction should be done by block averaging, which is slower but loses less detail than pixel sampling.
- Registration forces the features in more than one image of the same scene to be in the same pixel position.

3

Filters

Most images are affected to some extent by *noise*, that is unexplained variation in data: disturbances in image intensity that are either uninterpretable or not of interest. Image analysis is often simplified if this noise can be *filtered out*. In an analogous way, filters are used in chemistry to free liquids from suspended impurities by passing them through a layer of sand or charcoal. Engineers working in signal processing have extended the meaning of the term *filter* to include operations that accentuate features of interest in data. Employing this broader definition, image filters may be used to emphasize *edges*—that is, boundaries between objects or parts of objects in images. Filters provide an aid to visual interpretation of images, and can also be used as a precursor to further digital processing, such as segmentation (Chapter 4).

Most of the methods considered in Chapter 2 operated on each pixel separately. Filters change a pixel's value, taking into account the values of neighbouring pixels too. They may either be applied directly to recorded images, such as those in Chapter 1, or after transformation of pixel values as discussed in Chapter 2. To take a simple example, Figs 3.1(b–d) show the results of applying three filters to the cashmere fibres image, which has been redisplayed in Fig 3.1(a).

- Figure 3.1(b) is a display of the output from a 5×5 *moving average filter*. Each pixel has been replaced by the average of pixel values in a 5×5 square, or *window*, centred on that pixel. The result is to reduce noise in the image, but also to blur the edges of the fibres. A similar effect can be produced by looking at Fig. 3.1(a) through half-closed eyes.
- If the output from the moving average filter is subtracted from the original image, on a pixel-by-pixel basis, then the result is as shown in Fig. 3.1(c) (which has been displayed with the largest negative pixel values shown as black and the largest positive pixel values shown as white). This filter (the original image minus its smoothed version) is a *Laplacian filter*. It has had the effect of emphasizing edges in the image.
- Figure 3.1(d) shows the result produced when output from the Laplacian filter is added to the original image, again on a pixel-by-pixel basis. To the eye, this image looks clearer than Fig. 3.1(a) because transitions at edges have been magnified—an effect known as *unsharp masking*.

(a)

(b)

(c)

(d)

Fig. 3.1 Application of linear filters to cashmere image: (a) original image; (b) output from 5 × 5 moving average filter; (c) result of subtracting output of 5 × 5 moving average filter from original image; (d) result of adding original image to the difference between the output from 5 × 5 moving average filter and the original image.

We shall consider these three filters in more detail in §3.1.

The above filters are all linear, because output values are linear combinations of the pixels in the original image. Linear methods are far more amenable to mathematical analysis than are nonlinear ones, and are consequently far better understood. For example, if a linear filter is applied to the output from another linear filter then the result is a third linear filter. Also, the result would be the same if the order in which the two filters were applied was reversed. There are two, complementary, ways of studying linear filters, namely in the *spatial* and *frequency* domains. These approaches are considered in §§3.1 and 3.2 respectively. The less mathematical reader may prefer to skip §3.2. This can be done without losing the sense of the rest of the chapter.

Nonlinear filters—that is, all filters that are not linear—are more diverse and difficult to categorize, and are still an active area of research. They are potentially more powerful than linear filters because they are able to reduce noise levels without simultaneously blurring edges. However, their theoretical foundations are far less secure, and they can produce features that are entirely spurious. Therefore care must be taken in using them. In §3.3, some nonlinear smoothing filters are considered, and in §3.4, nonlinear edge-detection filters are introduced.

Finally, the key points of the chapter are summarized in §3.5.

3.1 LINEAR FILTERS IN THE SPATIAL DOMAIN

The moving average or *box filter*, which produced Fig. 3.1(b), is the simplest of all filters. It replaces each pixel by the average of pixel values in a square centred at that pixel. All linear filters work in the same way, except that, instead of forming a simple average, a weighted average is formed. Using the terminology of Chapter 1, let f_{ij}, for $i, j = 1, \ldots, n$, denote the pixel values in the image. We shall use g, with pixel values g_{ij}, to denote the output from the filter. A linear filter of size $(2m + 1) \times (2m + 1)$, with specified weights w_{kl} for $k, l = -m, \ldots, m$, gives

$$g_{ij} = \sum_{k=-m}^{m} \sum_{l=-m}^{m} w_{kl} f_{i+k, j+l} \quad \text{for} \quad i, j = (m+1), \ldots, (n-m).$$

For example, if $m = 1$ then the window over which averaging is carried out is 3×3, and

$$
\begin{aligned}
g_{ij} = \quad & w_{-1,-1} \quad f_{i-1,j-1} \quad +w_{-1,0} \quad f_{i-1,j} \quad +w_{-1,1} \quad f_{i-1,j+1} \\
& +w_{0,-1} \quad f_{i,j-1} \quad +w_{0,0} \quad f_{i,j} \quad +w_{0,1} \quad f_{i,j+1} \\
& +w_{1,-1} \quad f_{i+1,j-1} \quad +w_{1,0} \quad f_{i+1,j} \quad +w_{1,1} \quad f_{i+1,j+1}.
\end{aligned}
$$

For full generality, the weights w can depend on i and j, resulting in a filter that varies across the image. However, the linear filters considered in this chapter will all be spatially invariant. Also, all the filters will have windows composed of odd numbers of rows and columns. It is possible to have even-sized windows, but then there is a half-pixel displacement between the input and output images.

Note that the *borders* of g, that is

$$g_{ij} \quad \text{where either } i \text{ or } j = 1, \ldots, m \quad \text{or} \quad n - m + 1, \ldots, n,$$

have not been defined above. Various possibilities exist for dealing with them:

1. They could be discarded, resulting in g being smaller than f.
2. The pixels in the borders of g could be assigned the same values as those in the borders of f.

3. The border pixels in g could be set to zero.
4. The filter could be modified to handle incomplete neighbourhoods, for example

    (a) by ignoring those parts of the neighbourhood that lie outside the image;

    (b) by reflecting the input image f along its first and last row and column, so that pixel values $f_{i,n+1} = f_{i,n-1}$ etc.,

    (c) by wrapping-round the input image so that $f_{i,n+1} = f_{i,1}$ etc., as though it were on a torus.

In this chapter we shall take option 2 for smoothing filters and option 3 for edge-detection filters, except in §3.2, where we shall make use of option 4(c). Strictly speaking, this wrap-round approach is the only valid option for the mathematical results on linear filters to be applicable over the whole image.

If all the elements in w are *positive* then the effect of the filter is to smooth the image. The two most commonly used filters of this type, the moving average and the Gaussian, will be considered in §3.1.1. If some weights are *negative* then the filter outputs a difference between pixel values, which can have the effect of emphasising edges. Filters of this type will be presented in §3.1.2.

3.1.1 Smoothing

For the moving average filter, $w_{kl} = 1/(2m + 1)^2$. Figures 3.2(a), (c) and (e) show the results of applying moving average filters with windows of size 3×3, 5×5 and 9×9 to the transformed X-ray image in Fig. 2.3(b). As can be seen, the bigger the window the greater the noise reduction and blurring.

Computational efficiency is an important consideration in image analysis because of the size of data sets. In total, there are $(2m + 1)^2$ additions and multiplications per pixel involved in deriving g from f. However, some filters can be computed more quickly. A filter is said to be *separable* if it can be performed by first filtering the image inside a $(2m + 1) \times 1$ window, and then inside a $1 \times (2m + 1)$ window. In other words, it can be separated into a column operation

$$h_{ij} = \sum_{k=-m}^{m} w_k^c f_{i+k,j} \quad \text{for } i = m+1, \ldots, n-m; \quad j = 1, \ldots, n,$$

using column weights $w_{-m}^c, \ldots, w_m^c$, followed by a row operation:

$$g_{ij} = \sum_{l=-m}^{m} w_l^r h_{i,j+l} \quad \text{for } i, j = (m+1), \ldots, (n-m),$$

using row weights $w^r_{-m}, \ldots, w^r_m$. In order for this to be possible, the array of weights w_{kl} must be expressible as the product of the column and row weights, as follows:

$$w_{kl} = w^c_k w^r_l \quad \text{for } k, l = -m, \ldots, m.$$

The number of operations per pixel has been reduced from $(2m+1)^2$ to $2(2m+1)$. Therefore a separable filter can be computed more quickly than one that is not separable, even when $m = 1$.

Although the moving average filter is separable (with $w^c_k = w^r_l = 1/(2m+1)$), there exists a yet more efficient algorithm. This uses a recursive implementation, that is, one in which the output from the filter at location (i, j) is updated to obtain the output at location $(i+1, j)$. In contrast, the formulae we have considered so far involve calculating from scratch the output at each location. Specifically, the first $2m + 1$ pixel values in column j are averaged:

$$h_{m+1,j} = \frac{1}{2m+1} \sum_{k=1}^{2m+1} f_{kj}.$$

Then the pixel value in the first row (f_{1j}) is dropped from the average and the pixel in row $2m + 2$ is added. This operation is repeated for every value in column j, so that

$$h_{ij} = h_{i-1,j} + \frac{f_{i+m,j} - f_{i-m-1,j}}{2m+1} \quad \text{for } i = m+2, \ldots, n-m.$$

This procedure is repeated for each column $j = 1, \ldots, n$, to obtain h. Then the same algorithm is applied along each row of h, to obtain g. The number of operations per pixel has been reduced to 4 *irrespective of the filter size* (m).

Table 3.1 gives times for the general, separable and moving average algorithms considered above, implemented in a Fortran 77 program to run on a SUN Sparc2 computer. (Timings for filters to be discussed later in the chapter are also included.) Separable and, in particular, moving average filters run much more quickly than the general linear filter, particularly when image and window sizes are large.

Although the moving average filter is simple and fast, it has two drawbacks:

1. It is not *isotropic* (i.e. circularly symmetric), but smooths further along diagonals than along rows and columns.
2. Weights have an abrupt cut-off rather than decaying gradually to zero, which leaves discontinuities in the smoothed image.

Artefacts introduced by the square window can be seen in Fig. 3.2(e), particularly around the sheep's backbone. Drawback 1 could be overcome by calculating the average in a lattice approximation to a circular, rather than a

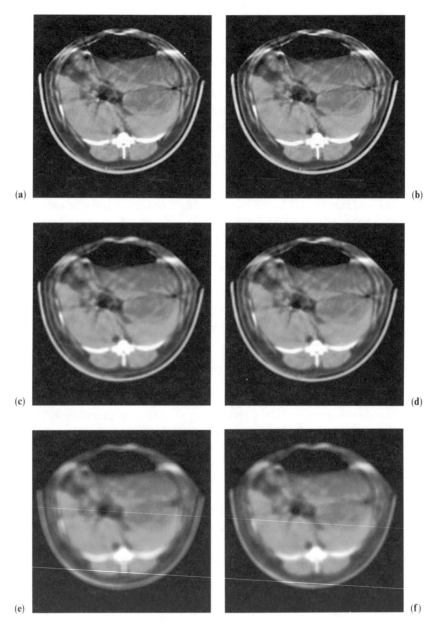

Fig. 3.2   Linear smoothing filters applied to X-ray image: (a) 3×3 moving average; (b) Gaussian, $\sigma^2 = \frac{2}{3}$; (c) 5×5 moving average; (d) Gaussian, $\sigma^2 = 2$; (e) 9×9 moving average; (f) Gaussian, $\sigma^2 = 6\frac{2}{3}$.

Table 3.1   Times for a SUN Sparc2 computer to process images of size $n \times n$, by applying a range of filters using neighbourhoods of size $(2m + 1) \times (2m + 1)$

		Computer time (tenths of a second)											
n		128				256				512			
m		1	2	3	4	1	2	3	4	1	2	3	4
Linear filters													
Moving average		1	1	1	1	4	4	4	4	16	16	16	16
Gaussian approximation		2	2	2	2	8	8	8	8	33	33	33	33
Separable filter		2	2	2	2	6	7	8	9	25	31	36	40
General linear filter		2	3	5	7	8	13	21	31	31	53	85	126
Fourier implementation		6	6	6	6	24	24	24	24	112	112	112	112
Nonlinear smoothing filters													
Median		1	1	1	2	4	5	5	6	13	18	22	26
Robust: fastest		2	2	2	2	7	8	9	10	25	32	38	43
slowest		4	4	4	5	12	15	17	19	48	61	71	79
Minimum variance		4	4	4	4	17	17	17	17	74	74	74	74
Nonlinear edge filters													
Roberts		1	—	—	—	2	—	—	—	10	—	—	—
Kirsch		17	—	—	—	70	—	—	—	287	—	—	—

square, neighbourhood. Such a filter with constant weights would not be separable, but could be implemented reasonably efficiently using a 2D version of the recursive algorithm.

Gaussian filters

Gaussian filters are the only ones that are separable and, at least to a lattice approximation, circularly symmetric. They also overcome the other stated drawback of moving average filters because weights decay to zero. Gaussian filters have weights specified by the probability density function of a bivariate Gaussian, or Normal, distribution with variance σ^2, that is

$$w_{ij} = \frac{1}{2\pi\sigma^2} \exp\left\{ \frac{-(i^2 + j^2)}{2\sigma^2} \right\} \quad \text{for } i, j = -[3\sigma], \ldots, [3\sigma],$$

for some specified positive value for σ^2. Here 'exp' denotes the exponential function and $[3\sigma]$ represents the 'integer part' of 3σ. Limits of $\pm 3\sigma$ are chosen because Gaussian weights are negligibly small beyond them. Note that the divisor of $2\pi\sigma^2$ ensures that the weights sum to unity (approximately), which is a common convention with smoothing filters. If $\sigma^2 = 1$, the array of weights is

$$w = \frac{1}{1000} \begin{pmatrix} 0 & 0 & 1 & 2 & 1 & 0 & 0 \\ 0 & 3 & 13 & 22 & 13 & 3 & 0 \\ 1 & 13 & 59 & 97 & 59 & 13 & 1 \\ 2 & 22 & 97 & 159 & 97 & 22 & 2 \\ 1 & 13 & 59 & 97 & 59 & 13 & 1 \\ 0 & 3 & 13 & 22 & 13 & 3 & 0 \\ 0 & 0 & 1 & 2 & 1 & 0 & 0 \end{pmatrix}.$$

(For succinctness of notation, we show a 7×7 array to represent the weights w_{kl} for $k, l = -3, \ldots, 3$, and we have specified the weights only to three decimal places—hence the divisor of 1000 at the beginning of the array.)

Figures 3.2(b), (d) and (f) show the results of applying Gaussian filters with $\sigma^2 = \frac{2}{3}, 2$ and $6\frac{2}{3}$ to the X-ray image. These three values of σ^2 give filters that average to the same extents (have the same variance) as the moving average filters already considered. Figures 3.2(a) and (b) can be seen to be very similar, as can Figs 3.2(c) and (d). However, the Gaussian filter has produced a smoother result in Fig. 3.2(f), when compared with the moving average filter in Fig. 3.2(e).

For certain values of σ^2, Gaussian filters can be implemented efficiently by approximating them by repeated applications of moving average filters. For example, in one dimension, four iterations of a moving average filter of width 3 is equivalent to filtering using weights of

$$\frac{1}{81}(0 \quad 1 \quad 4 \quad 10 \quad 16 \quad 19 \quad 16 \quad 10 \quad 4 \quad 1 \quad 0)$$

A Gaussian filter with a variance of $\sigma^2 = \frac{8}{3}$ has weights

$$\frac{1}{81}(0.2 \quad 1.0 \quad 3.7 \quad 9.4 \quad 16.4 \quad 19.8 \quad 16.4 \quad 9.4 \quad 3.7 \quad 1.0 \quad 0.2)$$

The agreement is seen to be very good. In general, four repeats of a moving average filter of size $(2m + 1) \times (2m + 1)$ approximates a Gaussian filter with $\sigma^2 = \frac{4}{3}(m^2 + m)$ (Wells, 1986). Therefore, in order to use the approximation, we must have $\sigma^2 \geq \frac{8}{3}$, for which the window is at least 9×9. If four iterations of a moving average filter are used, the number of operations per pixel is 16, irrespective of the value of σ^2. Table 3.1 gives computer timings for this approximation of a Gaussian filter. Provided m can be found such that $\frac{4}{3}(m^2 + m)$ is an acceptable value for σ^2, the iterated algorithm is faster than a separable implementation in a $(2[3\sigma] + 1) \times (2[3\sigma] + 1)$ window.

As an aside, it may be noted that the Gaussian filter can also be used to simultaneously smooth and interpolate between pixels, provided that $\sigma^2 \geq 1$. At location (y, x), for non-integer row index y and column index x, the estimated intensity is an average of local pixel values, weighted by the Gaussian density function:

$$g(y,x) = \sum_{i=[y-3\sigma]}^{i=[y+3\sigma]} \sum_{j=[x-3\sigma]}^{j=[x+3\sigma]} \frac{f_{ij}}{2\pi\sigma^2} \exp\left\{\frac{-\{(i-y)^2 + (j-x)^2\}}{2\sigma^2}\right\}.$$

This is an example of *kernel regression* (Eubank, 1988).

Any set of weights w_{ij} that are all positive will smooth an image. Alternative strategies for choosing weights include template matching, local fitting of polynomial surfaces, and minimum-error predictors. The latter case will be returned to in §3.2.3 as an example of a Wiener filter. Hastie and Tibshirani (1990, Ch. 2) review alternative statistical approaches to smoothing. As with all smoothing operations, there is the fundamental trade-off between *variance* and *bias*: a filter that operates in a larger neighbourhood will be more effective at reducing noise, but will also blur edges.

3.1.2 Edge detection

To emphasize edges in an image, it is necessary for some of the weights in the filter to be negative. In particular, sets of weights that sum to zero produce, as output, differences in pixel values in the input image. A *first-derivative row filter* gives, at a position in the output image, the difference in pixel values in columns on either side of that location in the input image. Therefore the output from the filter will be large in magnitude (either negative or positive) if there is a marked difference in pixel values to the left and right of a pixel location. One set of weights, which also involves some smoothing in averaging over 3 rows, is given by

$$w = \frac{1}{6}\begin{pmatrix} -1 & 0 & 1 \\ -1 & 0 & 1 \\ -1 & 0 & 1 \end{pmatrix}.$$

This choice of weights gives a separable filter, achieves some noise reduction and, in a sense to be explained below, estimates the first derivative of the image intensity in the row direction.

To show that the output does estimate the first derivative, consider the image intensity $f(y,x)$ to be specified for continuously varying row index y and column index x, and to be differentiable. By a Taylor series expansion, f near (i,j) can be approximated by f_{ij}, together with a sum of partial derivatives of f with respect to y and x evaluated at (i,j):

$$f_{i+k,j+l} \approx f_{ij} + k\frac{\partial f_{ij}}{\partial y} + l\frac{\partial f_{ij}}{\partial x} + \frac{k^2}{2}\frac{\partial^2 f_{ij}}{\partial y^2} + \frac{l^2}{2}\frac{\partial^2 f_{ij}}{\partial x^2} + kl\frac{\partial^2 f_{ij}}{\partial y \partial x}.$$

It is found, after some messy algebra in which most terms cancel, that for the

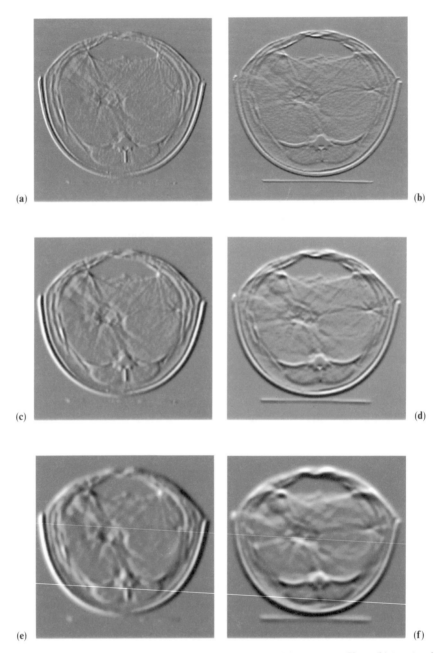

Fig. 3.3   First-derivative filters applied to X-ray image: (a) 3×3 row filter; (b) 3×3 column filter;   (c) row filter after Gaussian, $\sigma^2 = 2$; (d) column filter after Gaussian, $\sigma^2 = 2$; (e) row filter after Gaussian, $\sigma^2 = 6\frac{2}{3}$; (f) column filter after Gaussian, $\sigma^2 = 6\frac{2}{3}$.

above array of weights,

$$\sum_{k=-1}^{1}\sum_{l=-1}^{1} w_{kl}\, f_{i+k,\,j+l} \approx \frac{\partial f_{ij}}{\partial x},$$

i.e. a first derivative in the row direction.

Figure 3.3(a) shows the result when the first-derivative row filter is applied to the X-ray image, and Fig. 3.3(b) shows the column counterpart, a *first-derivative column filter* obtained using

$$w = \frac{1}{6}\begin{pmatrix} -1 & -1 & -1 \\ 0 & 0 & 0 \\ 1 & 1 & 1 \end{pmatrix}.$$

(These images have been displayed with zero pixel values shown as mid-grey, positive values as lighter greys and negative values as darker grey.) The results are strikingly similar to views of a landscape illuminated by a low sun. Our eyes cannot resist seeing the resultant image as 3D, even though it isn't. Note that the row filter produces non-zero values in response to edges that are approximately vertical (i.e. down columns of the image), whereas the column filter responds most to horizontal (along row) edges.

Although these filters do well at emphasizing edges, they have the disadvantage of also emphasizing noise, as can be seen in Figs 3.3(a, b). The effects of noise can be diminished by evaluating the filters over larger sizes of window. A simple way to achieve this is by applying the derivative filters to the output from a smoothing filter such as the Gaussian already considered. No further smoothing is necessary, so w can be simplified to a 1×3 filter

$$w = \frac{1}{2}(-1 \quad 0 \quad 1),$$

for the first-derivative row filter, and its transpose, a 3×1 filter

$$w = \frac{1}{2}\begin{pmatrix} -1 \\ 0 \\ 1 \end{pmatrix},$$

for the first-derivative column filter. The results of applying these filters to images, after Gaussian smoothing with $\sigma^2 = 2$ and $6\frac{2}{3}$ (i.e. the images displayed in Figs 3.2(d, f) are shown in Figs 3.3(c–f)). The combined filter has weights that are a convolution of the weights in the two constituent filters. In the case of the first-derivative row filter with $\sigma^2 = 2$, the weights are

$$w = \frac{1}{1000}\begin{pmatrix} 0 & -1 & -3 & -3 & 0 & 3 & 3 & 1 & 0 \\ -2 & -5 & -10 & -9 & 0 & 9 & 10 & 5 & 2 \\ -3 & -11 & -21 & -20 & 0 & 20 & 21 & 11 & 3 \\ -4 & -14 & -27 & -25 & 0 & 25 & 27 & 14 & 4 \\ -3 & -11 & -21 & -20 & 0 & 20 & 21 & 11 & 3 \\ -2 & -5 & -10 & -9 & 0 & 9 & 10 & 5 & 2 \\ 0 & -1 & -3 & -3 & 0 & 3 & 3 & 1 & 0 \end{pmatrix}.$$

Therefore the filter output at a pixel is again the difference in pixel values to the left and right of it, but here more values have been included in the averaging process on each side.

Linear combinations of first-derivative row and column filters can be used to produce first-derivative filters in other directions. But unfortunately they have to be combined in a nonlinear way in order to be sensitive to edges at all orientations simultaneously. Hence further discussion of them will be deferred until we consider nonlinear edge filters in §3.4.

Laplacian filters

Second-derivative filters, unlike first-derivative ones, can be isotropic, and therefore responsive to edges in any direction, while remaining linear. Consider the array of weights

$$w = \frac{1}{3}\begin{pmatrix} 1 & 1 & 1 \\ 1 & -8 & 1 \\ 1 & 1 & 1 \end{pmatrix}.$$

This is the same type of filter as that used to produce Fig. 3.1(c), i.e. the difference between the output from a moving average filter and the original image. It can be shown that, with these weights,

$$\sum_{k=-1}^{1} \sum_{l=-1}^{1} w_{kl}\, f_{i+k,j+l} \approx \frac{\partial^2 f_{ij}}{\partial x^2} + \frac{\partial^2 f_{ij}}{\partial y^2},$$

which is the *Laplacian* transform of f. Figure 3.4(a) shows the result of applying this filter to the X-ray image. (As with Fig. 3.3, zero pixel values are displayed as mid-grey.) Edges in the original image manifest themselves as *zero-crossings* in the Laplacian—that is, on one side of an edge the Laplacian will be positive while on the other side it will be negative. This is because an edge produces a maximum or minimum in the first derivative, and therefore it generates a zero value in the second derivative. Figure 3.4(b) provides an aid to identification of zero-crossings: negative values from the Laplacian filter are displayed as black and positive values as white. Therefore zero-crossings correspond to black/white boundaries. As with the first-derivative filters, noise has unfortunately been accentuated together with edges.

To reduce the effects of noise, the above filter or a simpler version of the Laplacian filter,

$$w = \begin{pmatrix} 0 & 1 & 0 \\ 1 & -4 & 1 \\ 0 & 1 & 0 \end{pmatrix},$$

can be applied to the image produced after Gaussian smoothing of the original. The results with $\sigma^2 = 2$ and $6\frac{2}{3}$ are shown in Figs 3.4(c) and (e), and zero-crossings are shown as the black/white boundaries in Figs 3.4(d) and (f). As

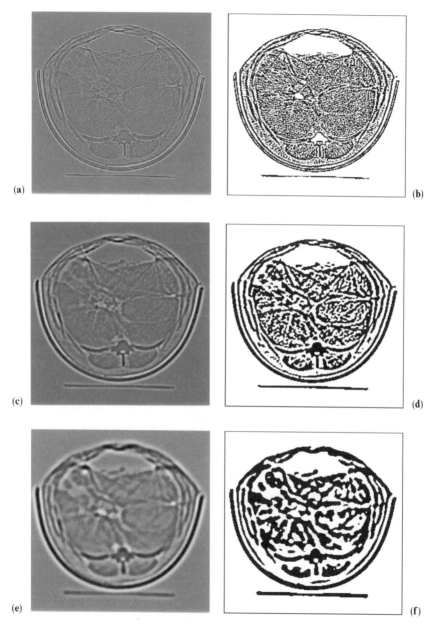

Fig. 3.4 Second-derivative filters applied to X-ray image: (a) 3×3 Laplacian filter; (b) zero-thresholded version of (a); (c) Laplacian-of-Gaussian filter, $\sigma^2 = 2$; (d) zero-thresholded version of (c); (e) Laplacian-of-Gaussian filter, $\sigma^2 = 6\frac{2}{3}$; (f) zero-thresholded version of (e).

σ^2 increases, the number of apparent edges decreases. These combined filters, obtained by Gaussian smoothing and then applying the Laplacian filter, are termed *Laplacian-of-Gaussian* filters. They were proposed by Marr and Hildreth (1980) in a seminal paper on vision. It is of interest to note that the weights, which have the expression

$$w_{ij} = \frac{1}{2\pi\sigma^6}(i^2 + j^2 - 2\sigma^2)\exp\left\{\frac{-(i^2 + j^2)}{2\sigma^2}\right\},$$

can be re-expressed as the derivative of the Gaussian weights with respect to σ^2:

$$2\frac{\partial}{\partial\sigma^2}\left(\frac{1}{2\pi\sigma^2}\exp\left\{\frac{-(i^2 + j^2)}{2\sigma^2}\right\}\right),$$

and so can be approximated by the difference between a pair of Gaussian filters with different values of σ^2, the so-called *difference-of-Gaussian (DoG) filter*. An image can be decomposed into a series of differences between outputs of Gaussian filters at a range of variances, to give a *multiresolution representation* of an image. The new field of wavelets extends these ideas and gives them a more rigorous foundation (see eg. Mallat, 1989).

All the filters we have considered so far have had weights that are either all positive, or which sum to zero. However, such restrictions are not necessary. For example, the *unsharp masking filter* that was used to produce Fig. 3.1(d) has weights

$$w = \frac{1}{9}\begin{pmatrix} -1 & -1 & -1 \\ -1 & 17 & -1 \\ -1 & -1 & -1 \end{pmatrix}.$$

This filter emphasizes edges, but (approximately) retains the original pixel values in edge-free regions of an image. It is used to aid visual interpretation of images rather than as an edge-detection filter.

3.2   LINEAR FILTERS IN THE FREQUENCY DOMAIN

Instead of representing an image as an $n \times n$ array of pixel values, we can alternatively represent it as the sum of many sine waves of different frequencies, amplitudes and directions. This is referred to as the *frequency domain* or *Fourier representation*. The parameters specifying the sine waves are termed *Fourier coefficients*. For some scientists, particularly engineers, this seems an obvious thing to do—for others, it may take some getting used to. The reasons why we are taking this approach are as follows:

• Extra insight can be gained into how linear filters work by studying them in the frequency domain.

- Some linear filters can be computed more efficiently in the frequency domain, by using the fast Fourier transform (FFT).
- New filters can be identified.

In §3.2.1 we shall present the basic theory of Fourier transforms. Then, in §3.2.2, we shall look again at the linear filters already considered in §3.1. Finally, in §3.2.3, we shall develop some new filters by specifying them in the frequency domain. The less-mathematical reader may prefer to skip the rest of this section, which can be done without losing the sense of the rest of the chapter.

3.2.1   Theory of the Fourier transform

The *Fourier transform* of f is the $n \times n$ array f^* defined by

$$f_{kl}^* = \sum_{i=1}^{n} \sum_{j=1}^{n} f_{ij} \exp\left\{\frac{-2\pi i}{n}(ik + jl)\right\} \quad \text{for } k, l = 1, \ldots, n,$$

where $i = \sqrt{-1}$. The inverse transform is very similar, but recovers f from f^*. It is

$$f_{ij} = \frac{1}{n^2} \sum_{k=1}^{n} \sum_{l=1}^{n} f_{kl}^* \exp\left\{\frac{2\pi i}{n}(ik + jl)\right\} \quad \text{for } i, j = 1, \ldots, n.$$

It is arbitrary where the divisor n^2 appears in the two equations above, and different authors make different choices. (It should be borne in mind, however, that this choice affects some of the equations that follow.)

Alternatively, we can avoid using i. If the real part of f^* (which is a complex array) is denoted by a and the imaginary part (that is the term involving i) is denoted by b then

$$a_{kl} = \sum_{i=1}^{n} \sum_{j=1}^{n} f_{ij} \cos\left\{\frac{2\pi}{n}(ik + jl)\right\}, \quad b_{kl} = -\sum_{i=1}^{n} \sum_{j=1}^{n} f_{ij} \sin\left\{\frac{2\pi}{n}(ik + jl)\right\},$$

and

$$f_{ij} = \frac{1}{n^2} \sum_{k=1}^{n} \sum_{l=1}^{n} \left(a_{kl} \cos\left\{\frac{2\pi}{n}(ik + jl)\right\} - b_{kl} \sin\left\{\frac{2\pi}{n}(ik + jl)\right\} \right)$$

$$= \frac{1}{n^2} \sum_{k=1}^{n} \sum_{l=1}^{n} r_{kl} \cos\left\{\frac{2\pi}{n}(ik + jl) + \theta_{kl}\right\},$$

where r_{kl} and θ_{kl} are the *amplitude* and *phase* of the Fourier transform at frequency (k, l): $r_{kl} = \sqrt{a_{kl}^2 + b_{kl}^2}$, $\theta_{kl} = \tan^{-1}(b_{kl}/a_{kl})$. (Here, and subsequently, 'tan^{-1}' must produce output over a range of 2π radians, to ensure that $\cos\theta_{kl} = a_{kl}/r_{kl}$ and $\sin\theta_{kl} = b_{kl}/r_{kl}$. The Fortran 77 and C functions 'atan2' will achieve this). Therefore, as we can see, f is a sum of sine (or cosine) terms.

Simple example

Let us take a simple example, composed of only a few sine terms, to see what this transformation looks like. If f is the 4×4 array

$$f = \begin{pmatrix} 2 & 5 & 2 & 7 \\ 1 & 6 & 3 & 6 \\ 2 & 7 & 2 & 5 \\ 3 & 6 & 1 & 6 \end{pmatrix}$$

then, by substituting numbers into the above equations, we obtain the result that the real and imaginary parts of f^* are

$$a = \begin{pmatrix} 0 & 0 & 0 & 0 \\ 0 & 0 & 0 & 0 \\ 0 & 0 & 0 & 0 \\ 0 & 32 & 0 & 64 \end{pmatrix}, \quad b = \begin{pmatrix} -8 & 0 & 0 & 0 \\ 0 & 0 & 0 & 0 \\ 0 & 0 & 8 & 0 \\ 0 & 0 & 0 & 0 \end{pmatrix}.$$

These can be re-expressed as amplitudes and phases of

$$r = \begin{pmatrix} 8 & 0 & 0 & 0 \\ 0 & 0 & 0 & 0 \\ 0 & 0 & 8 & 0 \\ 0 & 32 & 0 & 64 \end{pmatrix}, \quad \theta = \frac{\pi}{2} \begin{pmatrix} -1 & * & * & * \\ * & * & * & * \\ * & * & 1 & * \\ * & 0 & * & 0 \end{pmatrix},$$

where '$*$' denotes θ_{kl} that are undefined because $a_{kl} = b_{kl} = 0$. Therefore f is the sum of four cosines, namely for $i, j = 1, \ldots, 4$:

$$f_{ij} = \frac{1}{16} r_{11} \cos \left\{ \frac{1}{2}\pi(i+j) + \theta_{11} \right\} + \frac{1}{16} r_{33} \cos \left\{ \frac{1}{2}\pi(3i+3j) + \theta_{33} \right\}$$

$$+ \frac{1}{16} r_{42} \cos \left\{ \frac{1}{2}\pi(4i+2j) + \theta_{42} \right\} + \frac{1}{16} r_{44} \cos \left\{ \frac{1}{2}\pi(4i+4j) + \theta_{44} \right\}.$$

This works out as

$$f = \frac{1}{2} \begin{pmatrix} 0 & -1 & 0 & 1 \\ -1 & 0 & 1 & 0 \\ 0 & 1 & 0 & -1 \\ 1 & 0 & -1 & 0 \end{pmatrix} + \frac{1}{2} \begin{pmatrix} 0 & -1 & 0 & 1 \\ -1 & 0 & 1 & 0 \\ 0 & 1 & 0 & -1 \\ 1 & 0 & -1 & 0 \end{pmatrix}$$

$$+ 2 \begin{pmatrix} -1 & 1 & -1 & 1 \\ -1 & 1 & -1 & 1 \\ -1 & 1 & -1 & 1 \\ -1 & 1 & -1 & 1 \end{pmatrix} + 4 \begin{pmatrix} 1 & 1 & 1 & 1 \\ 1 & 1 & 1 & 1 \\ 1 & 1 & 1 & 1 \\ 1 & 1 & 1 & 1 \end{pmatrix}.$$

Note that the first two arrays are identical, and both represent a sinusoidal wave in a direction from top-left to bottom-right of the arrays. The third array

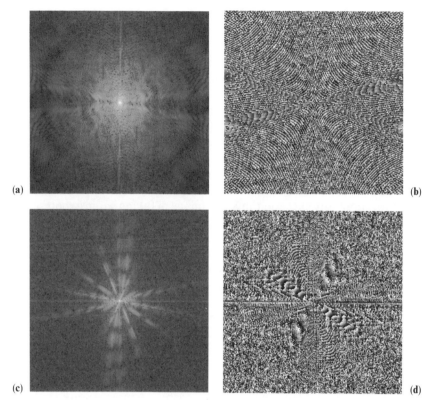

Fig. 3.5   Fourier transforms of images: (a) amplitudes of Fourier transform of X-ray image; (b) phases from X-ray image; (c) amplitudes of Fourier transform of cashmere image; (d) phases from cashmere image.

is a sinusoidal wave between columns, and at a higher frequency than the previous two, whereas the fourth array provides an overall mean.

Image examples

We shall now consider Fourier transforms of some images. Figures 3.5(a) and (b) show r and θ respectively for the X-ray image. Following common practice, in order to aid interpretation, these arrays have been displayed in the range $k, l = -\frac{1}{2}n, \dots, \frac{1}{2}n - 1$. Therefore the zero-frequency coefficient is positioned in the centres of the displays. (Because of the periodic nature of sines and cosines, the range of summations in the Fourier Transforms is arbitrary, subject to it being over an $n \times n$ square. Therefore $f_{kl}^* = f_{k+in,l+jn}^*$ for any values of i and j, and summation can range from $-\frac{1}{2}n$ to $\frac{1}{2}n - 1$.) The amplitude image has been displayed on a logarithmic scale, because otherwise all except the large values at low frequencies would appear as black. In the phase image, an angle of $-180°$ is displayed as black and one of $+180°$ is displayed as white. Note that r has $180°$ rotational symmetry and θ has $180°$ rotational antisymmetry (i.e. $r_{kl} = r_{-k,-l}$ and $\theta_{kl} = -\theta_{-k,-l}$) in these and all subsequent Fourier transforms. The amplitude display shows high coefficients at low frequencies, as is typical of images in general, whereas little can be discerned from the phase display.

Figures 3.5(c, d) show Fourier coefficients for the cashmere fibres image. The radiating spokes visible in Fig. 3.5(c) correspond to straight edges of fibres. Again, the amplitude shows high coefficients at low frequencies, and the phase display shows no discernible pattern.

To illustrate the information contained in r and θ separately, let us look at what happens to the images if r is modified before back-transforming to f. Figure 3.6(a) shows the effect on the X-ray image of taking square roots of all the Fourier amplitudes, and then applying the inverse transform to produce a new image. Figure 3.6(b) is the result of setting all amplitudes equal and then back-transforming. Therefore these images emphasize the information content of the phases *alone*. In both cases, edges have been emphasized in a way akin to Figs 3.1(c, d). (Note that the operations that produced Figs 3.6(a, b) are both nonlinear filters.) Contrary to the the impression given in Fig. 3.5, the information content is generally greater in θ than in r for Fourier transforms of images. Figure 3.6(c) further illustrates the point. It is the image produced when the cashmere fibre Fourier amplitudes are combined with the X-ray phases. The resemblance to the original X-ray image is much closer than to the cashmere image. Conversely, Fig. 3.6(d) has the X-ray amplitude and cashmere phases, and looks more like the cashmere image. Stark (1987, Chs 6–8) considers the problem of reconstructing an image when only phase information, or only amplitude information, is available, which occurs with certain sensors.

Efficient computation

Finally in this subsection on the theory of Fourier transforms we shall consider computational matters. The Fourier transform is separable. This, combined with the existence of a highly efficient algorithm, the *fast Fourier transform* (*FFT*) (see eg. Chatfield, 1989) leads to very fast implementations of the transform. The column transform

$$f'_{kj} = \sum_{i=1}^{n} f_{ij} \exp\left(\frac{-2\pi i}{n} ik\right) \quad \text{for } k = 1, \dots, n$$

can be carried out separately for each column j from 1 to n, using a real-only implementation of the FFT. The subsequent row transform

$$f^*_{kl} = \sum_{j=1}^{n} f'_{kj} \exp\left(\frac{-2\pi i}{n} jl\right) \quad \text{for } l = 1, \dots, n$$

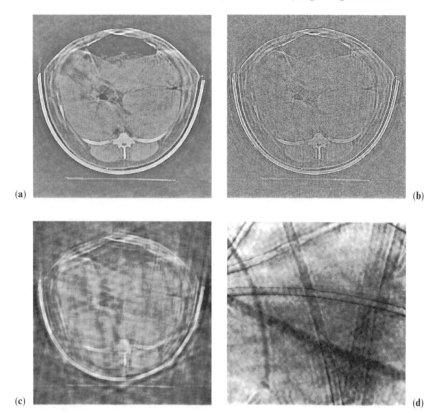

Fig. 3.6   Recovered images from Fourier domain after back-transforming: (a) X-ray image with Fourier amplitudes replaced by their square roots; (b) X-ray image with constant Fourier amplitudes; (c) cashmere amplitudes combined with X-ray phases; (d) X-ray amplitudes and cashmere phases.

requires a complex FFT algorithm (that is, one that operates on complex numbers), but need only be performed for $k = \frac{1}{2}n, \ldots, n$, because of the structure of f^*. The inverse transform similarly needs $\frac{1}{2}n + 1$ complex FFTs, but this time followed by n FFTs of Hermitian series, which are a restricted class of complex series. Numerical Algorithms Group routines CO6FPF, CO6FQF and CO6FRF (NAG, 1991) are Fortran 77 implementations of such FFTs, which work for any values of n. Yet more efficient FFT algorithms exist, which require n to be a power of 2 or to have only small prime factors. Also, hardware implementations of FFTs exist in some computers.

3.2.2   Filters already considered

Linear filters have a very simple representation in the frequency domain. We need to use a slightly different definition of a filter from that used in §3.1. We

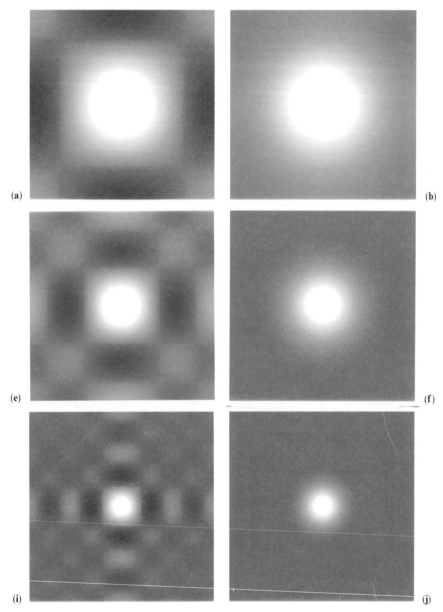

Fig. 3.7   Fourier transforms of linear filters used in Figs 3.2–3.4 (all transforms are real, except for the first-derivative filters, which are purely imaginary–in the latter case, the imaginary term is displayed): (a) 3×3 moving average; (b) Gaussian, $\sigma^2 = \frac{2}{3}$; (c) 3×3 first-derivative row filter; (d) Laplacian; (e) 5×5 moving average; (f) Gaussian, $\sigma^2 = 2$; (g) first-derivative row filter after Gaussian, $\sigma^2 = 2$; (h) Laplacian-of-Gaussian, $\sigma^2 = 2$; (i) 9×9 moving average; (j) Gaussian, $\sigma^2 = 6\frac{2}{3}$; (k) first-derivative row filter after Gaussian, $\sigma^2 = 6\frac{2}{3}$; (l) Laplacian-of-Gaussian, $\sigma^2 = 6\frac{2}{3}$.

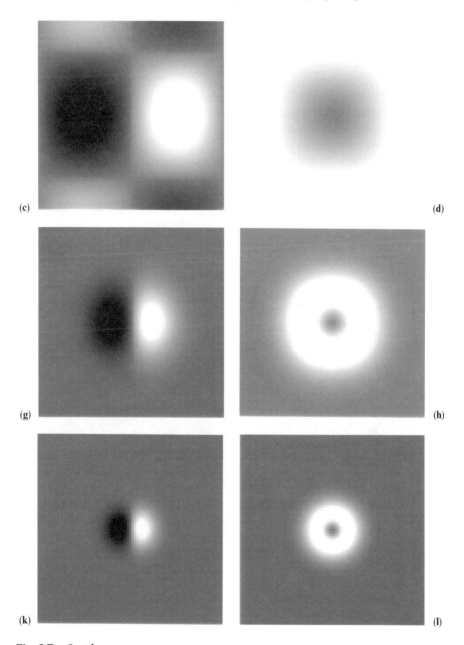

(c)

(d)

(g)

(h)

(k)

(l)

Fig. 3.7   *Contd.*

shall use

$$g_{ij} = \sum_{k=-(n/2)}^{(n/2)-1} \sum_{l=-(n/2)}^{(n/2)-1} w_{kl} \, f_{i+k,j+l} \quad \text{for } i,j = 1,\ldots,n,$$

where $i + k$ is replaced by $i + k \pm n$, if it falls outwith the range $1,\ldots,n$. A similar transformation is applied to $j + l$. In effect, the borders of f are wrapped round to meet each other as though the image lay on the surface of a torus, something we mentioned at the beginning of §3.1. Normally w is non-zero only for (k,l) near $(0,0)$, in which case the difference in results from those obtained using other border conditions will be slight. Remarkably, the Fourier transforms of f, g and w satisfy

$$g_{kl}^* = w_{kl}^* f_{kl}^* \quad \text{for } k,l = 1,\ldots,n,$$

provided that w and f are arrays of the same size. A filtering operation in the spatial domain has become simply a *point-by-point multiplication* in the frequency domain. Therefore an alternative way of applying a linear filter, instead of using the methods of §3.1, is to compute Fourier transforms of f and w, multiply their Fourier coefficients and back-transform to obtain g.

If w has $180°$ rotational symmetry (that is, $w_{kl} = w_{-k,-l}$), as have the moving average, Gaussian and Laplacian filters, then w^* is real. Therefore g^* has the same phases, θ, as f^*—only the amplitudes are different. Another possibility is that w is antisymmetric (that is, $w_{kl} = -w_{-k,-l}$), as were the first-derivative filters. In this case, w^* is purely imaginary. Such filters rescale the amplitudes and add or subtract $\frac{1}{2}\pi$ from the phases. Figure 3.7 shows w^* for all the filters used to produce Figs 3.2– 4. In the case of the first-derivative filters, the imaginary parts of w^* are shown. Zero is shown as mid-grey in each display.

The smoothing filters, that is the moving average and Gaussian, are *low-pass filters*, because these filters only let through low-frequency terms. All high-frequency terms in g^* are almost reduced to zero because w_{kl}^* decays to zero as (k,l) moves away from $(0,0)$. As the extent of smoothing is increased (i.e. as m or σ^2 is increased), the decay is more rapid; therefore progressively more of the higher frequencies are filtered out of the image. The Fourier transform of the array of moving average weights is approximately given by

$$w_{kl}^* = w_k^* w_l^*, \quad \text{where } w_k^* = \frac{\sin\{(2m+1)\pi k/n\}}{(2m+1)\pi k/n}.$$

It is not isotropic, and some elements in w are negative. (The expression for w_k^* is sometimes referred to as a 'sinc' function.) The Fourier transform of the Gaussian weights can be approximated by

$$w_{kl}^* = \exp\left\{\frac{-(k^2 + l^2)}{2(n/2\pi\sigma)^2}\right\}.$$

It is a scaled version of a Gaussian density, but with a different spread, and is both isotropic and non-negative.

The Laplacian filter, whose Fourier transform is shown in Fig. 3.7(d), is a *high-pass filter*, because it filters out only frequencies near zero. (Note that Fig. 3.7(d) is the complement of Fig. 3.7(a): element w_{kl}^* in one is equal to $1 - w_{kl}^*$ in the other.) The Laplacian-of-Gaussian filters (Figs 3.7h, l) are *band-pass filters* because they remove both low and high frequencies from an image. Therefore they operate as edge detectors, whilst managing to smooth out some of the noise. The first-derivative row filters are also band-pass filters, as can be seen in Figs 3.7(c, g, k).

For the case where w^* is known (and therefore w does not need Fourier transforming), the computer timings for implementing filters using Fourier transforms are given in Table 3.1. Such implementations would appear to be worthwhile only for filters that are non-separable and bigger than 7×7. Faster times are possible if n is a power of 2 and a specialized FFT algorithm is used. However, if w^* has itself to be obtained by applying an FFT to w then times will be slower. A further complication in making comparisons between spatial and frequency algorithms is that the former can often be achieved using integer arithmetic, while the latter always need floating-point operations, which are generally slower.

3.2.3   New filters and image restoration

So far in §3.2, we have only reviewed filters previously considered in §3.1. However, it is also possible to formulate filters in the frequency domain: a filter is specified by the array w^*, rather than by the array w. One class of isotropic filters, termed *'ideal' low-pass filters*, is given by setting w_{kl}^* to zero beyond a certain distance R from the origin, and otherwise setting it to unity:

$$w_{kl}^* = \begin{cases} 1 & \text{if } k^2 + l^2 < R^2, \\ 0 & \text{otherwise,} \end{cases} \quad \text{for } k, l = -\tfrac{1}{2}n, \dots, \tfrac{1}{2}n - 1.$$

'Ideal' high-pass filters are defined in a complementary way by setting w_{kl}^* to zero within a distance R from the origin:

$$w_{kl}^* = \begin{cases} 0 & \text{if } k^2 + l^2 < R^2, \\ 1 & \text{otherwise,} \end{cases} \quad \text{for } k, l = -\tfrac{1}{2}n, \dots, \tfrac{1}{2}n - 1.$$

Figure 3.8 shows the results when 'ideal' low-pass filters are applied to the X-ray image for $R = 60$, 30 and 15—values that approximately match the extents of smoothing in Fig. 3.2. A ripple or *'ringing effect'* is evident, particularly in Fig. 3.8(b), which can be explained by examining the weights w of the equivalent spatial filter. Because w has radial symmetry, we need only look at

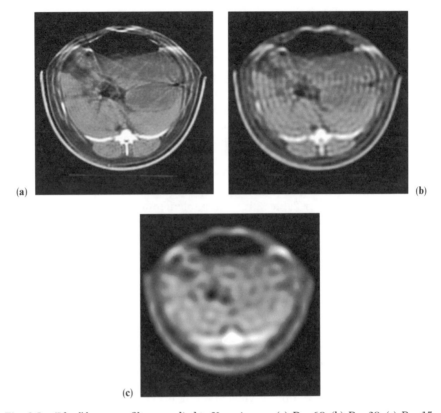

(a)   (b)

(c)

Fig. 3.8   'Ideal' low-pass filters applied to X-ray image: (a) $R = 60$; (b) $R = 30$; (c) $R = 15$.

elements in a single row. For $R = 60$.

$$
\begin{array}{rccccccccccc}
l = & 0 & 1 & 2 & 3 & 4 & 5 & 6 & 7 & 8 & 9 \\
1000 \times w_{0l} = & 175 & 131 & 41 & -17 & -16 & 5 & 10 & -1 & -6 & 0
\end{array}
$$

The negative weights are the cause of the ringing. It is the same phenomenon by which moving average filters produced ringing effects in the Fourier domain (see e.g. Fig. 3.7i). The Gaussian filter is in some ways a compromise between these filters—its weights remain positive, but decrease smoothly in both spatial and frequency domains. Wang, Vagnucci and Li (1983) survey alternative, smooth specifications of w^* .

Restoration

If an image has been contaminated by noise and blurring of known forms, then filters can be constructed which optimally (in some sense) restore the original image. Some authors regard restoration as being distinct from image filtering, and devote whole chapters to it (see eg. Rosenfeld and Kak, 1982; Gonzalez and Wintz, 1987). There are both linear and nonlinear restoration methods. Here we shall consider only the fundamental linear method, called the *Wiener filter*.

The observed image f is assumed to be a blurred version of an underlying 'true' image g, with noise added. We envisage that g is a clear image, unaffected by noise and with distinct boundaries between objects, whereas in f the edges are spread over several pixels, and there is noise. We can express the degradation process as

$$f_{ij} = \sum_{k=-(n/2)}^{(n/2)-1} \sum_{l=-(n/2)}^{(n/2)-1} v_{kl}\, g_{i+k,\,j+l} + e_{ij} \quad \text{for } i, j = 1, \ldots, n,$$

where v denotes the weights by which g is blurred and e denotes the noise.

We can use information about the nature of the degradations to design a filter that will smooth f and enhance the edges, so as to get as close as possible to restoring g. Provided that we can consider g to be the realization of a random process, and subject to various technical conditions, the Wiener filter is the best linear predictor (in the least-squares sense) of g from f. The restored image $\hat{g}$ has Fourier transform

$$\hat{g}_{kl}^* = \frac{f_{kl}^*}{v_{kl}^*}\; \frac{|v_{kl}^*|^2}{|v_{kl}^*|^2 + S_{kl}^e / S_{kl}^g} \quad \text{for } k, l = -\tfrac{1}{2}n, \ldots, \tfrac{1}{2}n - 1.$$

For a derivation, see e.g. Rosenfeld and Kak (1982, §7.3). S_{kl}^g denotes the *spectrum* of g at frequency (k, l), that is, the expected value of the squared amplitude of g_{kl}^*, after scaling:

$$S_{kl}^g = \frac{1}{n^2} E(|g_{kl}^*|^2).$$

Similarly, the array S^e denotes the spectrum of e. S^g can alternatively be expressed as the Fourier transform of the *autocovariance function* C^g of g; that is,

$$S^g = C^{g*},$$

where C_{kl}^g denotes the covariance between pixel values in g that are k rows and l columns apart:

$$C_{kl}^g = E\{(g_{ij} - \bar{g})(g_{i+k,\,j+l} - \bar{g})\}$$

and $\bar{g}$ denotes the mean pixel value in g (see e.g. Chatfield, 1989).

In the *absence of blur*, $v_{kl}^* = 1$ for all values of k and l, and the Wiener filter simplifies to

$$\hat{g}_{kl}^* = \frac{f_{kl}^*}{1 + S_{kl}^e / S_{kl}^g},$$

which is typically a low-pass filter, also known in the spatial domain as simple kriging (Cressie, 1991, p. 110). In the *absence of noise*, $S^e_{kl} = 0$ for all values of k and l, and the Wiener filter simplifies to a high-pass filter:

$$\hat{g}^*_{kl} = \frac{f^*_{kl}}{v^*_{kl}}$$

which reverses the blurring process. However, in practice, this filter *never works* because there are always some values of v^*_{kl} which are close to zero. The presence of noise means that the highest-frequency components of g cannot be recovered from f, and the filter is band-pass.

SAR application

To illustrate the low-pass case, consider the log-transformed SAR image (Fig. 2.6), with the mean subtracted. The speckle noise has known variance of 0.41 (§1.1.3) and zero autocovariances at other lags. Therefore the noise spectrum is

$$S^e_{kl} = 0.41 \quad \text{for all } k, l.$$

There is assumed to be no blur, so S^g can be obtained either by estimating $S^f - S^e$, or indirectly by estimating C^g, which is equal to $C^f - C^e$. We shall follow the latter approach. The sample autocovariances of f are given in Table 3.2. They were computed efficiently by taking the inverse Fourier transform of $|f^*|^2/n^2$. Once a value of 0.41 (due to C^e) has been subtracted at $(k, l) = (0, 0)$, the statistics are consistent with an isotropic model with autocovariance function:

$$C^g_{kl} = 0.75 \times 0.9^{\sqrt{k^2+l^2}}$$

Table 3.2 Sample covariances of log-transformed pixel values in SAR image, for a range of row and/or column separations.

				$100 \times$ sample covariances				
Row				Column				
	-4	-3	-2	-1	0	1	2	3	4
0					116	61	52	48	44
1	43	46	51	59	73	56	51	47	44
2	42	45	48	53	55	51	47	45	42
3	40	44	47	49	50	48	45	42	40
4	39	42	44	46	47	45	42	40	37

Fig. 3.9   Wiener filter applied to SAR image.

(Horgan, 1994). When these expressions are used in the Wiener filter, it simplifies to an isotropic spatial filter with weights

$$
\begin{array}{lcccc}
l = & 0 & 1 & 2 & 3 \\
1000 \times w_{0l} = & 204 & 77 & 24 & 6
\end{array}
$$

The result of applying the filter to the log-transformed SAR image is shown in Fig. 3.9. This filter is the best compromise in linear filters between a large window to smooth the speckle noise and a small window to avoid blurring edges such as field boundaries. It should be borne in mind, however, that, although the Wiener filter is the optimal linear one, nonlinear filters to be considered in §3.3 can out-perform it.

DNA application

To illustrate deconvolution, that is reversal of the blurring process, we shall consider the DNA image and adopt an approach similar to that of Press *et al.* (1992, pp. 539–542). Figure 3.10(a) shows a single column of the image (i.e. of f) after inversion of pixel values and subtraction of a linear background trend. Each band on the gel appears in this plot as a peak spread over several

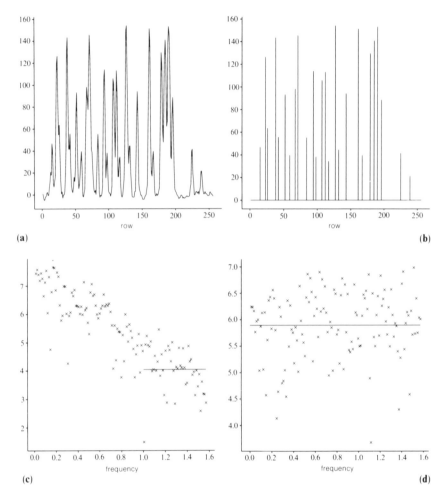

Fig. 3.10 Interpretation of single column of DNA image in order to identify degradation model required by Wiener filter: (a) pixel values after removal of linear trend and inverting greyscale; (b) idealized version of (a); (c) log-transformed amplitudes of Fourier transform of (a) plotted against frequency, together with mean value at higher frequencies; (d) log-transformed amplitudes of Fourier transform of (b), together with mean value; (e) phase differences between Fourier transforms of (a) and (b); (f) log-transformed ratios of amplitudes of Fourier transform of (a) and (b), together with estimated quadratic function; (g) result of applying Wiener filter to data in (a).

pixels, whereas the true pattern (g) is assumed to be distinct single-pixel-width spikes as shown in Fig. 3.10(b), which correspond to the local maxima in Fig. 3.10(a) above a threshold. The log-transformed amplitudes of the 1D Fourier transforms of Figs 3.10(a) and (b), that is $\log |f^*|$ and $\log |g^*|$, are shown plotted against frequency in Figs 3.10(c) and (d) respectively. Figure 3.10(e)

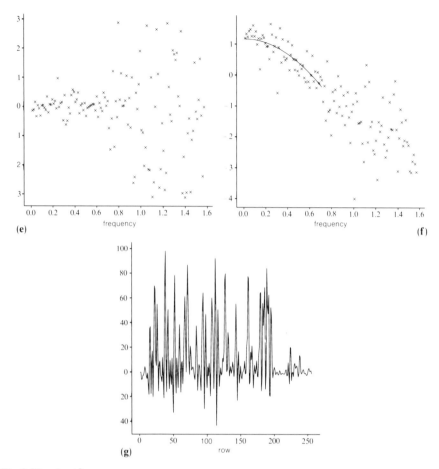

Fig 3.10 *Contd.*

shows the phase spectrum, which is the difference in phases between f^* and g^*, modulo 2π. Phase differences are small at frequencies below 0.7, but, above this limit, noise appears to dominate. The log-ratio, $\log(|f^*|/|g^*|)$, plotted in Fig. 3.10(f), can be used to estimate $|v^*|$ at frequencies below 0.7. On a log-scale a quadratic function fits well, with

$$\log|v_k^*| = 1.19 - 3.13\left(\frac{\pi k}{n}\right)^2 \qquad \text{for} \quad \frac{\pi k}{n} < 0.7.$$

The fitted curve is also displayed in Fig. 3.10(f). This corresponds to the blur v being Gaussian. From Fig. 3.10(d), the pattern underlying $|g^*|$ appears to be uniform, with a mean value of 5.89 on a log-scale, as plotted. We conclude that the spectrum of g is a constant:

$$S^g = \frac{1}{256}(e^{5.89})^2 = 510.$$

We shall assume, in the absence of any other information, that the noise is uncorrelated, so-called white noise, as was the SAR speckle noise. Its variance can be estimated from the high frequencies of f^*. The mean value of $\log|f^*|$, for frequencies greater than 1.0, is 4.05. This is plotted on

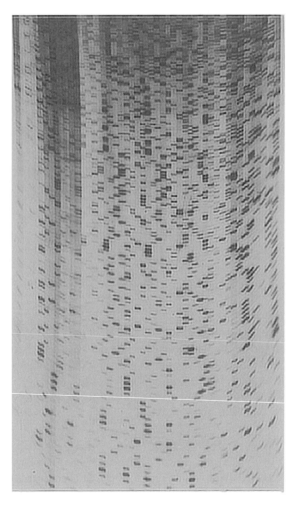

Fig. 3.11 Optimal linear deconvolution (Wiener filter) to restore DNA image.

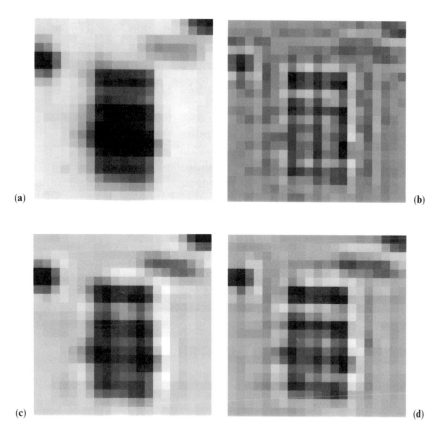

Fig. 3.12   Detail from DNA image: (a) original; (b) sub-optimal restoration assuming too little noise; (c) optimal restoration, as in Fig. 3.11; (d) sub-optimal restoration assuming too much noise.

Fig. 3.10(c). It leads to the spectrum of e being

$$S^e = 12.9.$$

Figure 3.10(g) shows the result of applying the Wiener filter to the data in Fig. 3.10(a), using the expressions identified above. The spread of the peaks has been reduced, but at the cost of a more variable signal.

If the 1D model is assumed to apply isotropically in 2D, the restored image after Wiener filtering is as given in Fig. 3.11. Figure 3.12 shows the effect on the restoration of a small part of the DNA image of varying the noise-to-signal ratio, S^e/S^g. Figure 3.12(a) shows the original image, in which four bands are just discernible. Figure 3.12(c) shows the optimal restoration, and the four bands appear clearly. Figure 3.12(b) shows the restoration if S^e/S^g is 10 times too small. Here, too many high frequencies have been used and noise dominates the result. Figure 3.12(d) shows the restoration with S^e/S^g 10 times too big. Not as much deblurring has been performed as is possible. For the optimal isotropic filter,

$$
\begin{array}{ccccccccccc}
l = & 0 & 1 & 2 & 3 & 4 & 5 & 6 & 7 & 8 \\
1000 \times w_{0l} = & 323 & 79 & -82 & 20 & 14 & -13 & 2 & 3 & -2
\end{array}
$$

Note that some weights are negative, as they have to be in a band-pass filter. These produce the ripple effects in the restored image, such as the white borders discernible around each band in Fig. 3.11.

Finally in this section, note that nonlinear restoration algorithms can do better than linear ones, but require substantially more computation. For example, maximum-entropy restoration (Skilling and Bryan, 1984) is one method that exploits the constraint that g is non-negative, which is appropriate to the DNA image after removal of background trend and grey-scale inversion. However, see also Donoho *et al.* (1992) for a discussion of alternative methods. It is also possible to apply deconvolution algorithms to three-dimensional images, to reduce blurring such as occurs in fluorescence microscopy (Shaw and Rawlins, 1991). These topics are all areas of active research, and further discussion of them is beyond the scope of this book.

3.3 NONLINEAR SMOOTHING FILTERS

In filtering to reduce noise levels, linear smoothing filters inevitably blur edges, because both edges and noise are high-frequency components of images. Nonlinear filters are able to *simultaneously* reduce noise and preserve edges. However,

- there are a bewildering number of filters from which to choose;
- they can be computationally expensive to use;
- they can generate spurious features and distort existing features in images.

Therefore they should be used with caution.

The simplest, most studied and most widely used nonlinear filter is the moving median. It will be considered in §3.3.1, together with similar filters based on the histogram of pixel values in a neighbourhood. In §3.3.2, a few of the many filters will be presented that make use of the spatial distribution of pixel values in a neighbourhood.

3.3.1 Histogram-based filters

The *moving median filter* is similar to the moving average filter, except that it produces as output at a pixel the median, rather than the mean, of the pixel values in a square window centred around that pixel. (The *median* of a set of numbers is the central one, once they have been sorted into ascending order.

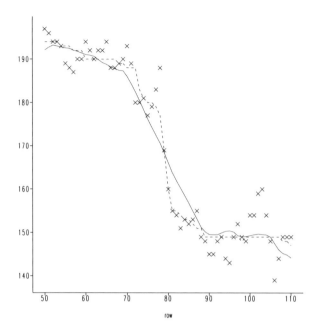

Fig. 3.13   Part of a row of pixels ($\times$) from muscle fibres image, together with moving average (———) and moving median (- - - - -) filters of width 19.

For example, the median of $\{1,0,4\}$ is 1, whereas the mean is $\frac{5}{3}$.) For a filter of size $(2m + 1) \times (2m + 1)$, the output is

$$g_{ij} = \text{median } \{f_{i+k,j+l} : k, l = -m, \ldots, m\} \quad \text{for } i, j = (m + 1), \ldots, (n - m).$$

Figure 3.13 shows a 1-D illustration of both moving average and moving median filters of width 19 applied to part of a row of pixels from the muscle fibres image. Both filters have reduced the noise level, the moving average more efficiently than the median, but the median filter has preserved the step-edge while the moving average filter has blurred it into a ramp of intermediate values.

An easily programmed way of computing the moving median filter would be to sort the set of pixel values $\{f_{i+k,j+l} : k, l = -m, \ldots, m\}$ into ascending order, then assign to g_{ij} the middle ranked value, and do this independently for every value of i and j from $(m + 1)$ to $(n - m)$. This approach is computationally slow and grossly inefficient. Huang, Yang and Tang (1979) presented a fast recursive algorithm for applying the median filter provided that the number of intensity levels of the image is not too great (no greater than 256 for example). It is based on a local histogram, which is updated from one neighbourhood of pixel values to the next and from which the median can be extracted without having to sort the set of $(2m + 1)^2$ pixel values. Their algorithm is as follows.

For each row of the image, $i = (m + 1), \ldots, (n - m)$, perform steps 1–9:

1. Set $j = m + 1$.
2. Calculate the histogram of $\{f_{i+k, j+l} : k, l = -m, \ldots, m\}$.
3. Set M = median value in the histogram
   and N = number of pixels in histogram that are less than M.
   (Note that M and N are simple to compute from the histogram.)
4. Increase j by 1.
5. Update the histogram
   to include $\{f_{i+k, j+m} : k = -m, \ldots, m\}$
   and exclude $\{f_{i+k, j-m-1} : k = -m, \ldots, m\}$.
   and simultaneously update N to remain equal to the number of pixels in
   the histogram that are less than M.
6. If $N > \frac{1}{2}(2m + 1)^2$
   then reduce M by 1, update N and repeat this step as many times as is
   necessary until the inequality is no longer satisfied; then go to step 8.
7. If N + (number of pixels equal to M) $< \frac{1}{2}(2m + 1)^2$
   then increase M by 1, update N and repeat this step as many times as are
   necessary until the inequality is no longer satisfied.
8. Set $g_{ij} = M$.
9. Return to step 4 if $j < n - m$.

Computer timings, given in Table 3.1, show that with this algorithm the median filter is as fast as linear filters to compute. (Note that it can be applied equally efficiently in a circular window, or any other convex region. The algorithm can also be adapted to find neighbourhood minima and maxima, which will be used in §3.4). Juhola, Katajainen and Raita (1991) review other algorithms, which are appropriate when there are more than 256 intensity levels.

Figures 3.14(a), (c) and (e) show the effects of median filters of sizes 3×3, 5×5 and 9×9 on the X-ray image. Larger windows produce more smoothing, and fine features are lost, but not to the same extent as with the linear filters in Fig. 3.2.

Nonlinear filters are not additive: repeated application of a median filter is not equivalent to a single application of a median filter using a different size of window. By making repeated use of a nonlinear smoothing filter with a small window, it is sometimes possible to improve on noise reduction while retaining fine details that would be lost if a larger window was used. Figures 3.14(b), (d) and (f) show the results after 5 iterations of the median filters corresponding to Figs 3.14(a), (c) and (e) respectively. Note the 'mottling effect' produced by regions of constant or near-constant pixel value, which are an inevitable consequence of median filtering and can lead to the identification of spurious boundaries.

Figure 3.15(a) shows 10 iterations of a 3×3 median filter applied to the log-transformed SAR image. Typically, each iteration produces less change in the image than the previous one until, after (potentially) many iterations, a stable

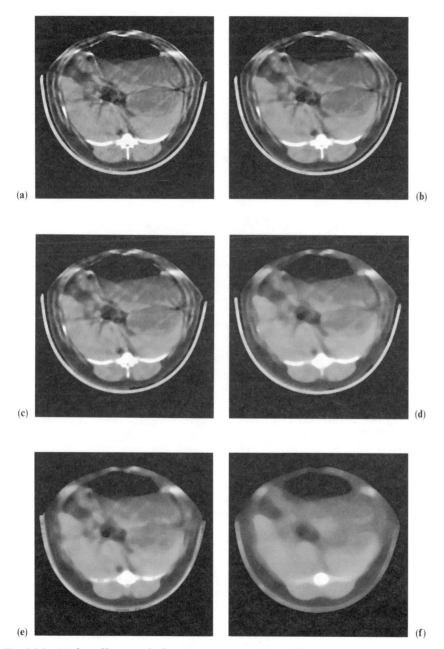

Fig 3.14 Median filters applied to X-ray image: (a) 3 × 3 filter; (b) 5 iterations of 3 × 3 filter; (c) 5 × 5 filter; (d) 5 iterations of 5 × 5 filter; (e) 9 × 9 filter; (f) 5 iterations of 9 × 9 filter.

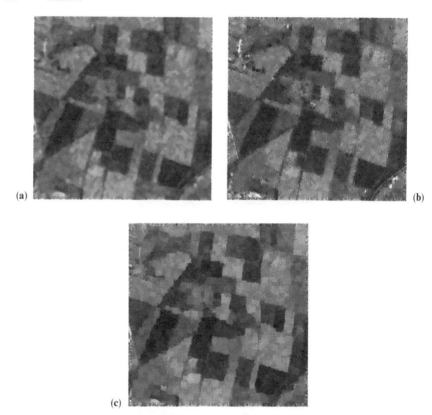

(a)   (b)   (c)

Fig. 3.15   Nonlinear smoothing filters applied to SAR image: (a) 10 iterations of 3×3 median filter; (b) 7×7 Lee's (1981) filter; (c) 9×9 minimum-variance filter with 5×5 sub-windows.

image is found that is unaffected by the median filter. However, iterated median filters have been found to produce artefacts with some images, and also to distort edges, so it is advisable to perform at most a few iterations.

In the past 30 years, statisticians have considered many robust estimators of the centre of a histogram, of which the median is but one. These estimators are robust in the sense that they are less affected than the mean by the presence of extreme values in the distribution. Moreover, the median has been shown to be outperformed by some other estimators, any of which could be used as an image filter thus:

$$g_{ij} = \text{robust estimator of centre of } \{f_{i+k,j+l} \; : \; k, l = -m, \ldots, m\}.$$

For example, a *trimmed mean* could be obtained by averaging the 50% of pixel values nearest to the median in each window. This should preserve edges, like the median, but should smooth more effectively. It can be programmed efficiently by making further use of the local histogram, using the kind of method described by Mitchell and Mashkit (1992). Because the speed of the algorithm is image dependent, it was run on all the 512×512 images used in the book. Table 3.1 reports the range of times taken by a SUN Sparc2 computer for a range of window sizes and $n \times n$ subsets of the full images. The muscle fibres image usually took longest and the Landsat band 2 took the least time. Such robust estimators, many of which are reviewed by Fong, Pomalaza-Raez and Wang (1989), are inevitably slower than the median filter. The computational burden is even further increased if iterations are involved. Also, the improvement over the median filter is often very small in practice.

3.3.2 Spatially adaptive filters

Median filters can be improved upon by making use, not only of the local histogram of values, but also of the spatial distribution of pixel values. Since this information is available it seems sensible to make use of it and not simply form a histogram.

One class of filters just makes a distinction between the *central pixel* and the rest of the window. For example, as a variant on the robust method in §3.3.1, the mean could be formed of the 50% of pixel values which are nearest f_{ij} in value. This is a case of k-neighbours filters, proposed by Davis and Rosenfeld (1978), and efficiently implemented by Mitchell and Mashkit (1992), in which in general the k pixel values nearest to f_{ij} are averaged. Computer times are almost identical to those for the robust method described in §3.3.1. Alternatively, pixels that differ from f_{ij} by less than the average intensity difference from f_{ij} could be averaged (Asano and Yokoya, 1981), or pixels could be weighted in inverse proportion to this difference (Wang *et al.*, 1981). If window size is varied, and filters are iterated, a vast number of filters may be generated.

If the noise variance (τ^2 say) is known, as with the SAR image, or can be estimated from homogeneous regions in an image, then use can be made of it. Lee (1983) proposed a filter in which smoothing is only performed where the output from a moving average filter is *consistent* with the original pixel value. Consistency is defined to be an output value within two standard deviations of f_{ij}. Specifically,

$$g_{ij} = \begin{cases} \bar{f}_{ij} & \text{if } |\bar{f}_{ij} - f_{ij}| < 2\tau, \\ f_{ij} & \text{otherwise,} \end{cases}$$

where $\bar{f}_{ij}$ denotes output from a moving average filter of size $(2m + 1) \times (2m + 1)$. In another paper, Lee (1981) proposed a variant on this idea, in which a weighted average is formed of f_{ij} and $\bar{f}_{ij}$:

$$g_{ij} = \bar{f}_{ij} + (f_{ij} - \bar{f}_{ij}) \frac{S_{ij}^2 - \tau^2}{S_{ij}^2},$$

where S_{ij}^2 is the sample variance in the window, that is,

$$S_{ij}^2 = \frac{1}{(2m+1)^2 - 1} \left\{ \sum_{k=-m}^{m} \sum_{l=-m}^{m} f_{i+k,j+l}^2 - (2m+1)^2 \bar{f}_{ij}^2 \right\},$$

provided this exceeds τ^2, and S_{ij}^2 is otherwise set equal to τ^2 in order to ensure that $S_{ij}^2 - \tau^2$ is non-negative. (Note that the summation term can be obtained efficiently by applying the moving average algorithm to an image with intensities f_{ij}^2.) Therefore, in an edge-free neighbourhood, $S_{ij}^2 \approx \tau^2$ and $g_{ij} \approx \bar{f}_{ij}$, whereas, near edges, $S_{ij}^2 \gg \tau^2$ and $g_{ij} \approx f_{ij}$. The filter is derived as the minimum mean-square-error predictor of g_{ij} if f_{ij} has mean g_{ij} and variance τ^2, and g_{ij} has mean $\bar{f}_{ij}$ and variance $S_{ij}^2 - \tau^2$. Durand, Gimonet and Perbos (1987) compared several filters for smoothing SAR images, and found this to be one of the best. The result is shown in Fig. 3.15(b) for a 7×7 window, which is the minimum size of window that can be used and still produce a reasonably precise estimate of S^2. Note that the centres of fields are smoothed, but the edges are left speckly. Also, bright spots on the top left and bottom right of the image remain unsmoothed, unlike in Figs 3.15(a, c).

The final class of filters we shall consider make greater use of the spatial distribution of pixels. Lev, Zucker and Rosenfeld (1977) proposed two sets of adaptive weights for a 3×3 window, which allow for the presence of edges and corners. Harwood *et al.* (1987) suggested forming a symmetric nearest-neighbours mean, by using the pixel value nearer f_{ij} for each pair of pixels symmetrically opposite in the window.

An elegant and fast method, originally proposed in a slightly different form by Tomita and Tsuji (1977), is the *minimum variance filter*. For this filter, the mean $\bar{f}$ and variance S are evaluated in five $(2m+1) \times (2m+1)$ subwindows within a $(4m+1) \times (4m+1)$ window, and the filter output is defined to be the mean of the subwindow that has the smallest variance. Therefore

$$g_{ij} = \bar{f}_{kl}, \quad \text{where } (k,l) = \text{argmin } \{S_{ij}^2, S_{i-m,j-m}^2, S_{i+m,j-m}^2, S_{i-m,j+m}^2, S_{i+m,j+m}^2\},$$

and 'argmin' means that $(k,l) = (i-m, j-m)$ if $S_{i-m,j-m}^2$ is the smallest of the five S^2s in the set, $(k,l) = (i+m, j-m)$ if $S_{i+m,j-m}^2$ is the smallest of the five S^2s, etc. The times taken by the computer to apply the filter are again given in Table 3.1. Times are unaffected by window size, because they make use of the moving average algorithm given in §3.1.1.

The output produced when the minimum-variance filter is applied to the SAR image, with $m = 2$, is shown in Fig. 3.15(c). Areas within fields have been smoothed effectively, although there appears to be some distortion of field boundaries. Square windows also have the drawback that they blur corners, that is, where edges meet at acute angles. Nagao and Matsuyama (1979) proposed a filter using essentially the same idea but with other shapes of window, and at the cost of a much more computationally intensive algorithm.

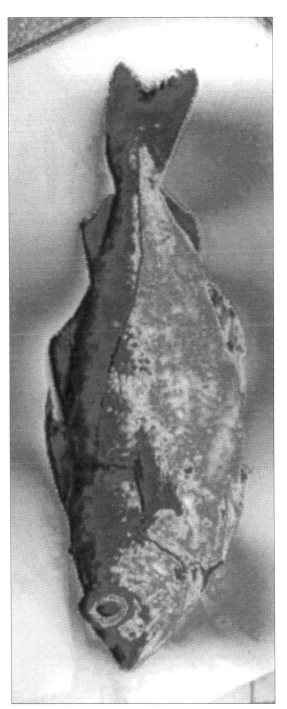

Fig. 2.9 Pseudocolour display of the fish image, using the map in Fig. 2.8(a).

(a)

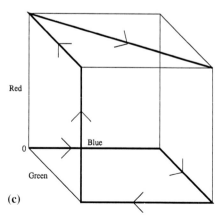

(c)

Fig. 2.8 Two pseudocolour maps: (a) is suitable for general use and (b) is appropriate when the data are angles or direction, since large values are displayed similarly to small values. (c) and (d) show the maps as paths in the colour cube. In (d), the black dot indicates the start and finish points of the path.

Red

0

Green

Blue

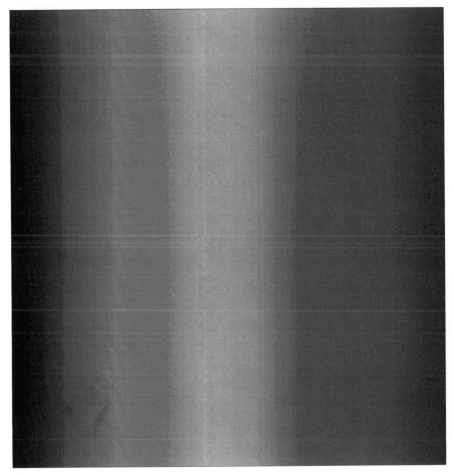

(b)

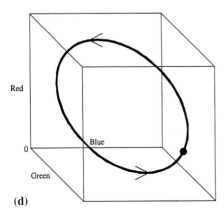

(d)

Fig. 2.8 *Contd.*

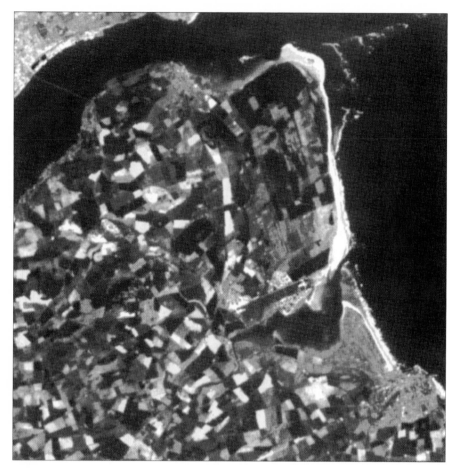

Fig. 2.10 Direct colour display of bands 1, 2 and 3 of the Landsat image.

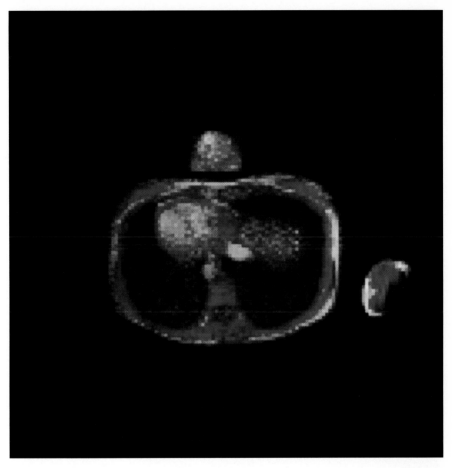

Fig. 2.11 Colour display of the MRI image variables. The inversion recovery variable controls the blue and green components of the display, while the proton density variable controls the red.

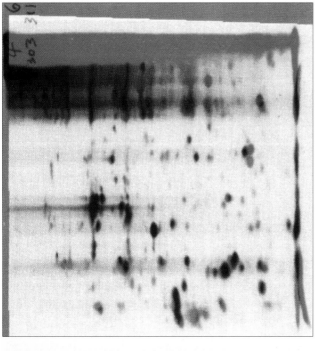

(b)

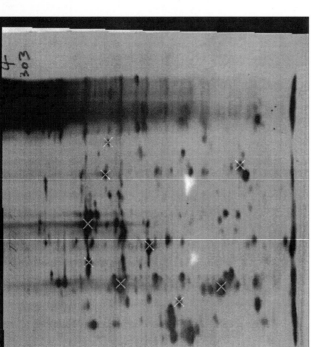

(a)

Fig. 2.17 Registration of electrophoresis image in Fig. 1.9(a) to Fig. 1.9(b). (a) shows the warped version of Fig. 1.9(a). The protein spots used as control points are shown by xs. (b) shows a superimposition of this image, in blue/green, with Fig. 1.9(b) in red. Spots that are in both images appear black, those in Fig. 1.9(a) only appear red and those in Fig. 1.9(b) only appear blue/green. This is because a black spot in the blue and green components that is not black in the red component appears red, and vice versa. This superimposition makes it easy to see protein spot variation.

Fig. 3.18 Prewitt's filter applied to DNA image, and displayed using intensity, hue and saturation: intensity is proportional to the square root of the maximum gradient, hue represents the direction of maximum gradient (modulo π), and saturation is set to be full everywhere.

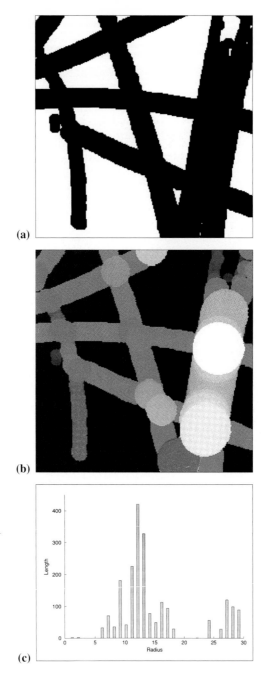

Fig. 5.18 Estimating fibre diameters: (a) image of Fig. 5.16(a) thresholded at a pixel value of 140; (b) sizing transform of (a), displayed with pseudocolour; (c) histogram of (b).

For descriptions of yet more filters, and comparative studies of performance, the interested reader is referred to the review papers of Chin and Yeh (1983), Wang *et al.* (1983), Mastin (1985), Fong *et al.* (1989), Imme (1991) and Wu, Wang and Liu (1992). Unfortunately, the reviews are ultimately inconclusive and contradictory, because images can be of such varying types and there is no unique measure of a filter's success.

3.4   NONLINEAR EDGE-DETECTION FILTERS

Edges are boundaries between objects, or parts of objects, in images. An edge filter should produce a large response at a location if pixels in the neighbourhood show a systematic pattern of changes in value. Linear filters, considered in §3.1.2, are of use as directional edge filters, but the Laplacian filter is the only one we have considered up to now that can detect edges at any orientation. In §3.4.1 we shall consider some simple, nonlinear filters for detecting edges in all directions, before taking a more systematic approach based on estimated derivatives in §3.4.2.

3.4.1   Simple edge filters

The *variance filter* S^2, already used in §3.3.2 in conjunction with Lee's filters, is a measure of edge presence that can be calculated very quickly. The result of its application to the X-ray image is shown in Fig. 3.16(a). The standard deviation rather than variance is displayed, with large values shown as darker pixels. As may be expected, neighbourhoods containing edges have larger variances, but the noise in the original image can also be seen in the filter output.

The *range filter* produces as output the range of pixel values in a window, that is the difference between the maximum and minimum values:

$$g_{ij} = \max \{f_{i+k,j+l} \; : \; k,l = -m, \ldots, m\} - \min \{f_{i+k,j+l} \; : \; k,l = -m, \ldots, m\}.$$

This is also a quick and simple filter, which can be obtained by modifying the moving median algorithm of §3.3.1. Alternatively, the filter can be implemented by making use of the property that the 'max' and 'min' filters are both separable (see §3.1.1). The result (Fig. 3.16b) is very similar to the variance filter for the X-ray image.

Roberts' filter is defined as the sum of the absolute differences of diagonally opposite pixel values:

$$g_{ij} = |f_{ij} - f_{i+1,j+1}| + |f_{i+1,j} - f_{i,j+1}|.$$

Because the window size is only 2×2, edges produce finer lines in the output, as illustrated in Fig. 3.16(c).

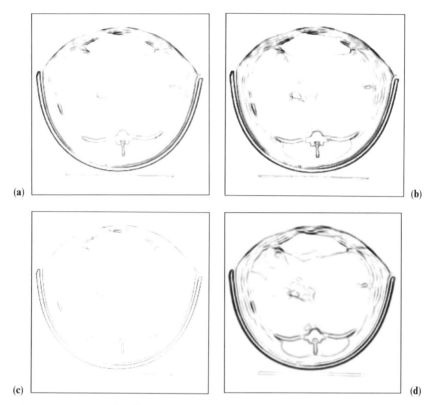

(a)   (b)

(c)   (d)

Fig. 3.16 Nonlinear edge filters applied to X-ray image: (a) square-root of variance filter; (b) range filter; (c) Roberts' filter; (d) Kirsch filter.

The *Kirsch filter* is one of a family of *template matching filters* (for others, see Jain, 1989, §9.4) in which the filter output is the maximum response from a set of linear filters that are sensitive to edges at different orientations:

$$g_{ij} = \max_{z=1,\dots,8} \sum_{k=-1}^{1} \sum_{l=-1}^{1} w_{kl}^{(z)} f_{i+k,j+l},$$

where

$$w^{(1)} = \begin{pmatrix} 5 & 5 & 5 \\ -3 & 0 & -3 \\ -3 & -3 & -3 \end{pmatrix}, \; w^{(2)} = \begin{pmatrix} 5 & 5 & -3 \\ 5 & 0 & -3 \\ -3 & -3 & -3 \end{pmatrix}, \; w^{(3)} = \begin{pmatrix} 5 & -3 & -3 \\ 5 & 0 & -3 \\ 5 & -3 & -3 \end{pmatrix},$$

and each subsequent weights matrix, $w^{(4)}, \dots, w^{(8)}$ has elements that are successively rotated through multiples of $45°$. This is a much slower algorithm than the preceding three, as can be seen from the computer times in Table 3.1. However, the results are usually very similar, as in Fig. 3.16(d).

All the above filters emphasize edges and noise simultaneously. To reduce the effect of noise, we need filters that also smooth. They could be obtained by applying the preceding filters to smoothed versions of images, such as those produced in §3.1.1 and §3.3, but a more elegant approach is motivated by estimation of gradients.

3.4.2 Gradient filters

Returning to the calculus notation of §3.1.2, the maximum gradient at a point (i, j) is given by

$$\sqrt{\left(\frac{\partial f_{ij}}{\partial x}\right)^2 + \left(\frac{\partial f_{ij}}{\partial y}\right)^2}.$$

We can design a filter to estimate this gradient by replacing the partial derivatives above by estimates of them: the outputs from first-derivative row and column filters from §3.1.2,

$$\widehat{\frac{\partial f_{ij}}{\partial x}} = \tfrac{1}{6}(f_{i-1,j+1} + f_{i,j+1} + f_{i+1,j+1} - f_{i-1,j-1} - f_{i,j-1} - f_{i+1,j-1})$$

and

$$\widehat{\frac{\partial f_{ij}}{\partial y}} = \tfrac{1}{6}(f_{i+1,j-1} + f_{i+1,j} + f_{i+1,j+1} - f_{i-1,j-1} - f_{i-1,j} - f_{i-1,j+1}).$$

This estimate of the maximum gradient is known as *Prewitt's filter*. Figure 3.17(a) shows the result of applying the filter to the X-ray image, which is equivalent to combining the images displayed in Figs 3.3(a) and (b), pixel-by-pixel, by taking the square root of the sum of their squares. Larger pixel values of the filter output are shown darker in the display, and zero values are shown as white. Note that this filter, unlike the first-derivative ones considered in § 3.1.2, responds to edges at all orientations.

Sobel's filter is very similar, except that in estimating the maximum gradient it gives more weight to the pixels nearest to (i, j), as follows:

$$\widehat{\frac{\partial f_{ij}}{\partial x}} = \tfrac{1}{8}(f_{i-1,j+1} + 2f_{i,j+1} + f_{i+1,j+1} - f_{i-1,j-1} - 2f_{i,j-1} - f_{i+1,j-1})$$

and

$$\widehat{\frac{\partial f_{ij}}{\partial y}} = \tfrac{1}{8}(f_{i+1,j-1} + 2f_{i+1,j} + f_{i+1,j+1} - f_{i-1,j-1} - 2f_{i-1,j} - f_{i-1,j+1}).$$

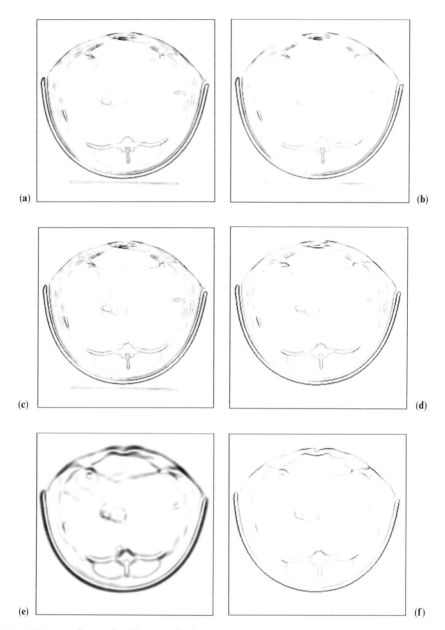

Fig. 3.17 Gradient edge filters applied to X-ray image: (a) Prewitt's filter; (b) (a) restricted to zero-crossings of directional second derivative; (c) maximum gradient after Gaussian, $\sigma^2 = 2$; (d) (c) restricted to zero-crossings; (e) maximum gradient after Gaussian, $\sigma^2 = 6\frac{2}{3}$; (f) (e) restricted to zero crossings.

For both Prewitt's and Sobel's filters, the direction of maximum gradient can also be obtained as

$$\phi_{ij} = \tan^{-1}\left(\widehat{\frac{\partial f_{ij}}{\partial y}} \bigg/ \widehat{\frac{\partial f_{ij}}{\partial x}}\right),$$

measured clockwise from horizontal in the direction of increasing x. Figure 3.18(colour plate) shows a colour display of the result of applying Prewitt's filter to the DNA image, using a hue–intensity–saturation transformation (see §2.3.3): orientation (modulo π) is displayed as hue, the maximum gradient is assigned to image intensity and saturation is set to full. It can be seen that reddish colours predominate on the left side of the display, where edges are vertical, and green–blue for edges of bands that are oriented from top-left to bottom-right. Horizontal bands generate blue colours in the centre of the display, whereas on the right side magenta predominates, as a result of bands being oriented from bottom-left to top-right. Glasbey and Wright (1994) use the orientation information to 'unwarp' the image of the gel so that all bands are horizontal.

Canny filter

To increase the amount of smoothing in the filters, derivatives can be estimated after Gaussian filtering, as in §3.1.2. Figure 3.17(c) shows the result for $\sigma^2 = 2$, obtained by combining the estimated first-derivatives displayed in Figs 3.3(c) and (d). Similarly, Fig. 3.17(e) shows the estimated gradient after Gaussian smoothing with $\sigma^2 = 6\frac{2}{3}$, derived from the data in Figs 3.3(e, f).

In an important paper, Canny (1986) shows that such filters are almost-optimal edge detectors. To overcome the problem of edge responses broadening in extent as smoothing increases, he suggested that positions of edges could be determined by finding zero-crossings of second derivatives in the direction of steepest gradient, that is, of

$$\cos^2 \phi_{ij} \frac{\partial^2 f_{ij}}{\partial x^2} + \sin^2 \phi_{ij} \frac{\partial^2 f_{ij}}{\partial y^2} + 2 \sin \phi_{ij} \cos \phi_{ij} \frac{\partial^2 f_{ij}}{\partial x\, \partial y},$$

where ϕ_{ij} is defined above. The directional second derivative can be estimated using an adaptive weights matrix:

$$w = \frac{1}{6}\begin{pmatrix} \frac{1}{2}\sin \phi_{ij} \cos \phi_{ij} & \sin^2 \phi_{ij} & -\frac{1}{2}\sin \phi_{ij} \cos \phi_{ij} \\ \cos^2 \phi_{ij} & -2 & \cos^2 \phi_{ij} \\ -\frac{1}{2}\sin \phi_{ij} \cos \phi_{ij} & \sin^2 \phi_{ij} & \frac{1}{2}\sin \phi_{ij} \cos \phi_{ij} \end{pmatrix}.$$

The effect is to thin the edges produced by the gradient filter. Figures 3.17(b), (d) and (f) show thinned versions of Figs 3.17(a), (c) and (e), in which all pixels have been set to zero except those in which the directional second-derivative changes sign between adjacent pixels.

The result is very similar to the Laplacian filter, shown in Fig. 3.4. However, the Laplacian is the sum of the directional second derivative and the second derivative perpendicular to the direction of maximum slope. Because the perpendicular derivative is dominated by noise, Canny claims that the directional derivative is a better edge detector than the Laplacian. For further substantial papers on the theoretical underpinning of edge filters, see Haralick (1984) and Torre and Poggio (1986).

3.5 SUMMARY

The key points about filters are as follows:

- Filters re-evaluate the value of every pixel in an image. For a particular pixel, the new value is based of pixel values in a local neighbourhood, a window centred on that pixel, in order to
  — reduce noise by smoothing, and/or
  — enhance edges.
- Filters provide an aid to visual interpretation of images, and can also be used as a precursor to further digital processing, such as segmentation (Chapter 4).
- Filters may either be applied directly to recorded images, such as those in Chapter 1, or after transformation of pixel values as discussed in Chapter 2.
- Filters are linear if the output values are linear combinations of the pixels in the original image, otherwise they are nonlinear.
  — Linear filters are well understood and fast to compute, but are incapable of smoothing without simultaneously blurring edges.
  — Nonlinear filters can smooth without blurring edges and can detect edges at all orientations simultaneously, but have less secure theoretical foundations and can be slow to compute.
  The main points about linear filters are as follows:
- Two smoothing filters are
  — the moving average filter;
  — the Gaussian filter.
- And four filters for edge detection are
  — first-derivative row and column filters;
  — first-derivative row and column filters combined with the Gaussian filter;
  — the Laplacian filter;
  — the Laplacian-of-Gaussian filter.
- Linear filters can be studied in the frequency domain, as well as in the spatial domain. The benefits are
  — efficient computation using the fast Fourier transform algorithm;

— further insight into how filters work, such as categorizing them as low-pass, high-pass or band-pass;

— opportunities to design new filters, such as
  • 'ideal' low-pass and high-pass filters, and
  • the Wiener filter for image restoration.

The main points about nonlinear filters are as follows:

• Five smoothing filters are
  — the moving median filter;
  — the trimmed-mean filter, as an example of robust estimation;
  — the k-neighbours filter;
  — Lee's filters;
  — the minimum-variance filter.

• Seven filters for edge detection are
  — the variance filter;
  — the range filter;
  — Roberts' filter;
  — Kirsch's template filter;
  — Prewitt's gradient filter;
  — Sobel's gradient filter;
  — the gradient filter combined with the Gaussian filter, and thinned using Canny's method.

The output from filters in this chapter will be used in Chapter 4 to segment images.

4

Segmentation

Image segmentation is the division of an image into *regions* or *categories*, which correspond to different objects or parts of objects. *Every* pixel in an image is allocated to one of a number of these categories. A good segmentation is typically one in which

- pixels in the same category have similar greyscale or multivariate values and form a connected region;
- neighbouring pixels that are in different categories have dissimilar values.

For example, in the muscle fibres image (Fig. 1.6b), each cross-sectional fibre could be viewed as a distinct object, and a successful segmentation would form a separate group of pixels corresponding to each fibre. Similarly, in the SAR image (Fig. 1.8g), each field could be regarded as a separate category.

Segmentation is often the critical step in image analysis: the point at which we move from considering each pixel as a unit of observation to working with *objects* (or parts of objects) in the image, composed of many pixels. If segmentation is done well then all other stages in image analysis are made simpler. But, as we shall see, success is often only partial when automatic segmentation algorithms are used. However, manual intervention can usually overcome these problems, and by this stage the computer should already have done most of the work.

This chapter fits into the structure of the book as follows. Segmentation algorithms may either be applied to the images as originally recorded (Chapter 1), or after the application of transformations and filters considered in Chapters 2 and 3. After segmentation, methods of mathematical morphology can be used to improve the results. These will be considered in Chapter 5. Finally, in Chapter 6, the segmentation results will be used to extract quantitative information from the images.

There are three general approaches to segmentation, termed thresholding, edge-based methods and region-based methods.

- In *thresholding*, pixels are allocated to categories according to the range of values in which a pixel lies. Figure 4.1(a) shows boundaries obtained by thresholding the muscle fibres image. Pixels with values less than 128 have been placed in one category, and the rest have been placed in the other category. The boundaries between adjacent pixels in different categories have been superimposed in white on the original image. It can be seen that the threshold has successfully segmented the image into the two predominant fibre types.

- In *edge-based* segmentation, an edge filter is applied to the image, pixels are classified as *edge* or *non-edge* depending on the filter output, and pixels that are not separated by an edge are allocated to the same category. Figure 4.1(b) shows the boundaries of connected regions after applying Prewitt's filter (§3.4.2) and eliminating all non-border segments containing fewer than 500 pixels. (More details will be given in §4.2.)

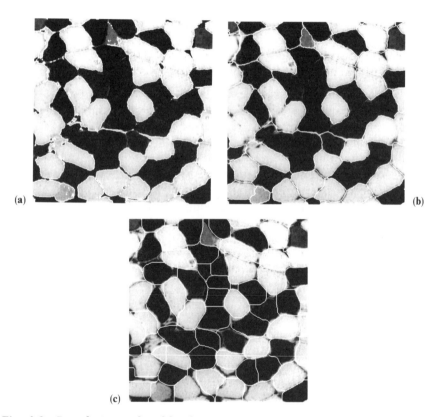

(a)   (b)

(c)

Fig. 4.1 Boundaries produced by three segmentations of the muscle fibres image: (a) by thresholding; (b) connected regions after thresholding the output of Prewitt's edge filter and removing small regions; (c) result produced by watershed algorithm on output from a variance filter with Gaussian weights ($\sigma^2 = 96$).

• Finally, *region-based* segmentation algorithms operate iteratively by grouping together pixels that are neighbours and have similar values, and splitting groups of pixels that are dissimilar in value. Figure 4.1(c) shows the boundaries produced by one such algorithm, based on the concept of *watersheds*, about which we will give more details in §4.3.

Note that none of the three methods illustrated in Fig. 4.1 has been completely successful in segmenting the muscle fibres image by placing a boundary between every adjacent pair of fibres. Each method has distinctive faults. For example, in Fig. 4.1(a), boundaries are well placed, but others are missing. In Fig. 4.1(c), however, more boundaries are present, and they are smooth, but they are not always in exactly the right positions.

The following three sections will consider these three approaches in more detail. Algorithms will be considered that can either be *fully automatic* or require some *manual intervention*. The key points of the chapter will be summarized in §4.4.

4.1 THRESHOLDING

Thresholding is the simplest and most commonly used method of segmentation. Given a single *threshold t*, the pixel located at lattice position (i, j), with greyscale value f_{ij}, is allocated to category 1 if

$$f_{ij} \leq t.$$

Otherwise, the pixel is allocated to category 2.

In many cases, t is chosen *manually* by the scientist, by trying a range of values of t and seeing which one works best at identifying the objects of interest. Figure 4.2 shows some segmentations of the soil image. In this application, the aim was to isolate soil material from the air-filled pores, which appear as the darker pixels in Fig. 1.6(c). Thresholds of 7, 10, 13, 20, 29 and 38 were chosen in Fig. 4.2(a–f) respectively to identify approximately 10, 20, 30, 40, 50 and 60% of the pixels as being pores. Figure 4.2(d), with a threshold of 20, looks best because most of the connected pore network evident in Fig. 1.6(c) has been correctly identified, as has most of the soil matrix.

Note the following:

• Although pixels in a single thresholded category will have similar values (either in the range 0 to t, or in the range $t + 1$ to 255), they will not usually constitute a single connected component. This is not a problem in the soil image, because the object (air) is not necessarily connected, either in the imaging plane or in three dimensions. In other cases, thresholding would be followed by dividing the initial categories into subcategories of connected regions.

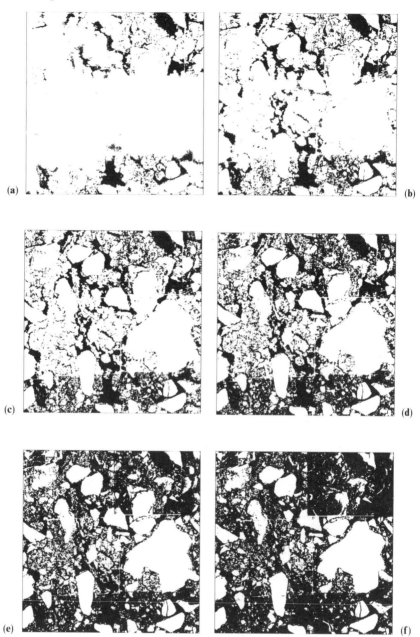

Fig. 4.2   Six segmentations of the soil image, obtained using manually selected thresholds of (a) 7, (b) 10, (c) 13, (d) 20, (e) 29 and (f) 38. These correspond to approximately 10%, 20%, . . . , 60% respectively of the image being displayed as black.

- More than one threshold can be used, in which case more than two categories are produced.
- Thresholds can be chosen *automatically*.

In §4.1.1, we shall consider algorithms for choosing the threshold on the basis of the histogram of greyscale pixel values. In §4.1.2, manually and automatically selected classifiers for multivariate images will be considered. Finally, in §4.1.3, thresholding algorithms that make use of context (that is, values of neighbouring pixels as well as the histogram of pixel values) will be presented.

4.1.1 Histogram-based thresholding

We shall denote the *histogram* of pixel values by $h_0, h_1, \ldots, h_N$, where h_k specifies the number of pixels in an image with greyscale value k and N is the maximum pixel value (typically 255). Ridler and Calvard (1978) and Trussell (1979) proposed a simple algorithm for choosing a single threshold. We shall refer to it as the *intermeans algorithm*. First we shall describe the algorithm in words, and then mathematically.

Initially, a guess has to be made at a possible value for the threshold. From this, the mean values of pixels in the two categories produced using this threshold are calculated. The threshold is repositioned to lie exactly half way between the two means. Mean values are calculated again and a new threshold is obtained, and so on until the threshold stops changing value. Mathematically, the algorithm can be specified as follows:

1. Make an initial guess at t; for example, set it equal to the median pixel value, that is, the value for which

$$\sum_{k=0}^{t} h_k \geq \frac{1}{2} n^2 > \sum_{k=0}^{t-1} h_k, \quad .$$

   where n^2 is the number of pixels in the $n \times n$ image.
2. Calculate the mean pixel value in each category. For values less than or equal to t, this is given by

$$\mu_1 = \sum_{k=0}^{t} k h_k \Bigg/ \sum_{k=0}^{t} h_k,$$

   whereas, for values greater than t, it is given by

$$\mu_2 = \sum_{k=t+1}^{N} k h_k \Bigg/ \sum_{k=t+1}^{N} h_k.$$

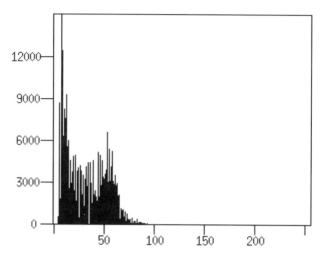

Fig. 4.3   Histogram of soil image.

3. Re-estimate t as half way between the two means, i.e.

$$t = \left[\frac{\mu_1 + \mu_2}{2}\right],$$

where [ ] denotes 'the integer part of' the expression between the brackets.

4. Repeat steps 2 and 3 until t stops changing value between consecutive evaluations.

Figure 4.3 shows the histogram of the soil image. From an initial value of $t = 28$ (the median pixel value), the algorithm changed t to 31, 32 and 33 on the first three iterations, and then t remained unchanged. The pixel means in the two categories are 15.4 and 52.3. Figure 4.4(a) shows the result of using this threshold. Note that this value of t is considerably higher than the threshold value of 20 that we favoured in the manual approach.

The intermeans algorithm has a tendency to find a threshold that divides the histogram in two, so that there are approximately equal numbers of pixels in the two categories. In many applications, such as the soil image, this is not appropriate. One way to overcome this drawback is to modify the algorithm as follows.

Consider a distribution that is a mixture of two Gaussian distributions. Therefore, in the absence of sampling variability, the histogram is given by

$$h_k = n^2\{p_1\phi_1(k) + p_2\phi_2(k)\} \qquad \text{for } k = 0, 1, \dots, N.$$

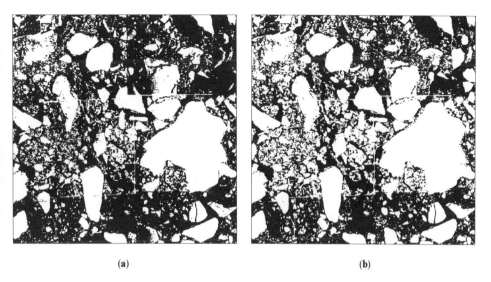

(a)   (b)

Fig. 4.4   Segmentations of the soil image, obtained by thresholding at (a) 33, selected by
the intermeans algorithm, and (b) 24, selected by the minimum-error algorithm.

Here p_1 and p_2 are proportions (such that $p_1 + p_2 = 1$), and $\phi_l(k)$ denotes the
probability density of a Gaussian distribution; that is,

$$\phi_l(k) = \frac{1}{\sqrt{2\pi\sigma_l^2}} \exp\left\{\frac{-(k - \mu_l)^2}{2\sigma_l^2}\right\} \qquad \text{for } l = 1, 2,$$

where μ_l and σ_l^2 are the mean and variance of pixel values in category l. The
best classification criterion, i.e. the one that misclassifies the least number of
pixels, allocates pixels with value k to category 1 if

$$p_1\phi_1(k) \geq p_2\phi_2(k),$$

and otherwise classifies them as 2. After substituting for ϕ and taking logs, the
inequality becomes

$$k^2\left(\frac{1}{\sigma_1^2} - \frac{1}{\sigma_2^2}\right) - 2k\left(\frac{\mu_1}{\sigma_1^2} - \frac{\mu_2}{\sigma_2^2}\right) + \left(\frac{\mu_1^2}{\sigma_1^2} - \frac{\mu_2^2}{\sigma_2^2} + \log\frac{\sigma_1^2 p_2^2}{\sigma_2^2 p_1^2}\right) \leq 0.$$

The left side of the inequality is a quadratic function in k. Let A, B and
C respectively denote the three terms in brackets. Then the criterion for

allocating pixels with value k to category 1 is

$$k^2 A - 2kB + C \leq 0.$$

There are three cases to consider:

(a) If $A = 0$ (i.e. $\sigma_1^2 = \sigma_2^2$), the criterion simplifies to one of allocating pixels with value k to category 1 if

$$2kB \geq C.$$

(If, in addition, $p_1 = p_2$ and $\mu_1 < \mu_2$, the criterion becomes $k \leq \frac{1}{2}(\mu_1 + \mu_2)$. Note that this is the intermeans criterion, which implicitly assumes that the two categories are of equal size.)

(b) If $B^2 < AC$ then the quadratic function has no real roots, and all pixels are classified as 1 if $A < 0$ (i.e. $\sigma_1^2 > \sigma_2^2$), or as 2 if $A > 0$.

(c) Otherwise, denote the roots t_1 and t_2, where $t_1 \leq t_2$ and

$$t_1, t_2 = \frac{B \pm \sqrt{B^2 - AC}}{A}.$$

The criteria for category 1 are

$$t_1 < k \leq t_2 \qquad \text{if } A > 0,$$
$$k \leq t_1 \text{ or } k > t_2 \qquad \text{if } A < 0.$$

In practice, cases (a) and (b) occur infrequently, and if $\mu_1 < \mu_2$ then the rule simplifies to the threshold:

$$\text{category 1 if a pixel value } k \leq \frac{B + \sqrt{B^2 - AC}}{A}.$$

Minimum-error algorithm

Kittler and Illingworth (1986) proposed an iterative *minimum-error algorithm*, which is based on this threshold and can be regarded as a generalization of the intermeans algorithm. Again, we will describe the algorithm in words, and then mathematically.

From an initial guess at the threshold, the proportions, means and variances of pixel values in the two categories are calculated. The threshold is repositioned according to the above criterion, and proportions, means and variances are recalculated. These steps are repeated until there are no changes in values between iterations. Mathematically, we have the following:

1. Make an initial guess at a value for t.
2. Estimate p_1, μ_1 and σ_1^2 for pixels with values less than or equal to t, by

$$p_1 = \frac{1}{n^2} \sum_{k=0}^{t} h_k,$$

$$\mu_1 = \frac{1}{n^2 p_1} \sum_{k=0}^{t} k h_k,$$

and

$$\sigma_1^2 = \frac{1}{n^2 p_1} \sum_{k=0}^{t} k^2 h_k - \mu_1^2.$$

Similarly, estimate p_2, μ_2 and σ_2^2 for pixels in the range $t+1, \ldots, N$.

3. Re-estimate t by

$$t = \left[\frac{B + \sqrt{B^2 - AC}}{A} \right],$$

where A, B, C and [ ] have already been defined.

4. Repeat steps 2 and 3 until t converges to a stable value.

When applied to the soil image, the algorithm converged in four iterations to $t = 24$. Figure 4.4(b) shows the result, which is more satisfactory than that produced by the intermeans algorithm because it has allowed for a smaller proportion of air pixels ($p_1 = 0.45$, compared with $p_2 = 0.55$). The algorithm has also taken account of the air pixels being less variable in value than those for the soil matrix ($\sigma_1^2 = 30$, whereas $\sigma_2^2 = 186$). This is in accord with the left-most peak in the histogram plot (Fig. 4.3) being quite narrow.

Note the following:

- The estimators in step 2 of the minimum-error algorithm are biased because they do not allow for overlaps between the distributions in the mixture, although Cho, Haralick and Yi (1989) have suggested one way of doing so. However, distributions of pixel values seldom conform exactly to the Gaussian, so such refinements seem unnecessary.
- Many other automatic, histogram-based thresholding algorithms have been proposed (for a review, see Glasbey, 1993). It was a popular area for research in the 1980s. Most algorithms have less theoretical justification than the two considered here.
- The intermeans and minimum-error algorithms extend simply to multiple thresholds, and to multivariate thresholds (which are the topic of §4.1.2). It is also possible to specify a different threshold in each part of the image (see e.g. Chow and Kaneko, 1972). Such an approach might have been of assistance in segmenting the soil image, because some of the electron

micrographs in the composite image montage had a greater overall brightness than others. Alternatively, trend can sometimes be removed before thresholding an image, for example by using a top-hat transform (§ 5.5).

4.1.2 Multivariate classifiers

There are many ways to extend the concept of a threshold in order to segment a multivariate image, such as the following:

- Allocate pixel (i, j) to category 1 if

$$f_{ij,m} \leq t_m \qquad \text{for } m = 1, \dots, M,$$

  where subscript m denotes the variate number, and there are M variates.
- Or, more generally, use a *box classifier* where the conditions for category 1 are

$$t_{1,m} < f_{ij,m} \leq t_{2,m} \qquad \text{for } m = 1, \dots, M.$$

- The condition could be a linear threshold:

$$c^T f_{ij} \leq t,$$

  (using *vector notation*, with $f_{ij} = (f_{ij,1}, f_{ij,2}, \dots, f_{ij,M})^T$, bold type used to denote vectors and the superscript T used to indicate a vector transpose).
- Or, the range of pixel values for category 1 could be a general set S in M-dimensional space:

$$f_{ij} \in S \subset \mathbb{R}^M.$$

For example, if $M = 2$, S could be a circular disc, the set of all integer pairs (k, l) such that $k^2 + l^2 < 100$.

For the rest of this subsection, we shall be making use of *matrix notation*. Less-mathematical readers may prefer to skip to §4.1.3, which can be done without losing the sense of the rest of the chapter.

Linear discrimination

Multivariate thresholding criteria are more difficult to specify than univariate ones. Therefore the approach often adopted in a segmentation that is manual (i.e. under the user's control) is to choose pixels which are known to belong to target categories (known as the *training set*) and then use them to classify the rest of the image. This is called *supervised classification*. Suppose that regions $B_1, \dots, B_R$ of the image have been identified as belonging to categories $1, \dots, R$ respectively, then the mean pixel values in the regions can be estimated as

$$\mu_r = \frac{1}{N_r} \sum_{(i,j) \in B_r} \sum f_{ij} \quad \text{for } r = 1, \ldots, R,$$

where summation is over all the N_r pixels in region B_r. The simplest case to consider is where the variance–covariance matrices in the different categories are assumed to be the same. In this case, the common variance–covariance matrix V can be estimated as

$$V = \sum_{r=1}^{R} \left(\sum_{(i,j) \in B_r} \sum f_{ij} f_{ij}^{T} - N_r \mu_r \mu_r^{T} \right) \bigg/ \sum_{r=1}^{R} N_r.$$

Distances in multidimensional space are measured taking into account the variance–covariance matrix V. The *squared Mahalanobis distance* between pixel value f_{ij} and the rth category mean is

$$(f_{ij} - \mu_r)^{T} V^{-1} (f_{ij} - \mu_r),$$

where the superscript '-1' indicates a matrix inverse.

The segmentation rule is to allocate each pixel in the image to the nearest category mean, as measured by the above distance. The rule simplifies to a set of linear thresholds. For example, in the two-category case, a pixel is allocated to category 1 if

$$\mu_1^{T} V^{-1} \mu_1 - 2\mu_1^{T} V^{-1} f_{ij} \leq \mu_2^{T} V^{-1} \mu_2 - 2\mu_2^{T} V^{-1} f_{ij}.$$

This is *linear discrimination*. If the variances are not assumed to be equal, then this results in *quadratic discrimination*. For more details, see e.g. Krzanowski (1988, §12.2).

Figure 4.5 shows this method applied to the bivariate image obtained by combining the MRI proton density and inversion recovery images of a woman's chest (Figs 1.7a, b). Four training areas have been selected manually as representatives of

1. lung tissue;
2. blood in the heart;
3. muscle, and other lean tissues;
4. adipose tissues, such as subcutaneous fat.

These areas are shown superimposed on the inversion recovery and proton density images in Fig. 4.5(a) and (b).

Figure 4.5(c) is a scatter plot of $f_{ij,2}$ against $f_{ij,1}$ for the training pixels, together with the linear separators of all pixel values into the four classes. For

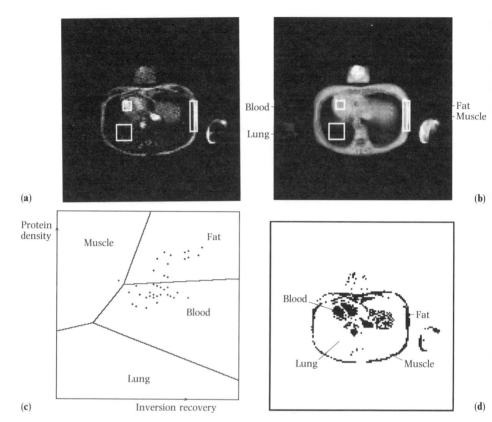

Fig. 4.5 Manual segmentation of the bivariate MRI image: (a) inversion recovery image with boundaries of four training areas superimposed—these are, from left to right, lung tissue, moving blood, muscle and subcutaneous fat; (b) proton density image with the same training areas shown; (c) scatter plot of values of training pixels, and linear discrimination classification, with points displayed in increasing levels of darkness for lung pixels, muscle, blood and fat; (d) segmentation of image by linear discrimination, using the same greyscale as in (c).

example, the muscle pixels are placed in the top-left of Fig. 4.5(c) because they have high values of proton density but low values from the inversion recovery signal, as can be seen in Figs 4.5(a, b). The points have been coded in progressively darker shades of grey, to denote lung, muscle, blood and fat pixels respectively. There is substantial overlap between blood and fat pixels in Fig. 4.5(c), so that even the training set has not been classified precisely.

Figure 4.5(d) shows the pixel allocation of the whole image. Air pixels, for which $f_{ij} = 0$, have been excluded from all classes. The lungs, shown as the lightest shade of grey, have been correctly identified, although note that boundary pixels around the whole body have also been misspecified in this category. Most other pixels have been classified as muscle, the second lightest shade of grey. The remaining pixels, displayed as dark grey for the blood category and black for fat, are mixed up, but taken together they serve to identify the major blood vessels and subcutaneous fat round the body.

k-means clustering

Automatic methods can be devised that do not require a training set. Such methods are termed *unsupervised classification*. One approach, *k-means clustering*, is an automatic version of linear discrimination. We shall describe the algorithm in words, and then more formally.

The number of categories, R, must be known. More commonly, several different values are tried. From an initial guess of the parameter values, that is the category means and the variance–covariance matrix, all pixels in an image are classified by the linear discrimination criterion. Then the parameters are estimated using all pixels classified in each category, and pixels are reclassified. These steps are repeated until there are no changes in values between iterations. More formally, we have the following:

1. Make initial guesses at $\mu_1, \ldots, \mu_R$, such as modes in the multivariate histogram or random values, and initially set the matrix V to the identity matrix.
2. Segment the image according to the linear discrimination criterion.
3. Estimate μ_r as the mean of pixel values classified as r, for $r = 1, \ldots, R$, and estimate V as the variance–covariance matrix averaged over categories, using the equations already given.
4. Repeat steps 2 and 3 until convergence.

Figure 4.6 shows the results produced by applying the *k*-means clustering algorithm to the bivariate MRI image. For values of R of 2, 3 and 4, the algorithm was run 50 times from randomly chosen starting points, and the segmentation that had the smallest within-group sum of squares,

$$\sum_{i=1}^{n}\sum_{j=1}^{n}\left\{ \min_{r=1,\ldots,R}\ (f_{ij} - \mu_r)^{\mathrm{T}} V^{-1}(f_{ij} - \mu_r)\right\},$$

was chosen. Figure 4.6(a) shows the bivariate histogram of all pixel values, the means of the two categories and the division between them, when two categories were sought. Figure 4.6(b) shows the segmented image using this criterion. Lung tissue appears to have been distinguished from other tissues. Figures 4.6(c) and (d) give analogous results for three clusters, and Fig. 4.6(e) and (f) for four clusters. The segmentation seems to be as good as that achieved in Fig. 4.5(d). The second-lightest shade of grey in Fig. 4.6(f) appears to identify muscle tissue, and blood and subcutaneous fat constitute a combined third and fourth class.

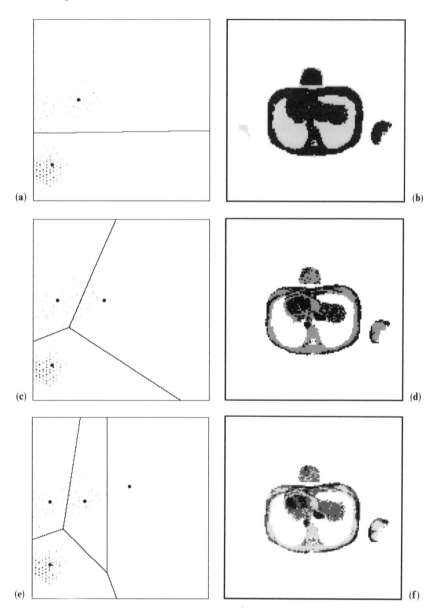

Fig. 4.6 Automatic segmentations of the MRI image into two, three and four classes: (a) bivariate histogram of pixel values, together with centres of two clusters and the division between them obtained by *k*-means clustering; (b) result displayed as an image; (c) bivariate histogram and optimal division into three clusters, and (d) image display; (e) bivariate histogram and optimal division into four clusters, and (f) image display.

The algorithm can be extended to handle unequal variances and proportions, as considered by Maronna and Jacovkis (1974). This is a multivariate generalization of the minimum error algorithm considered in §4.1.1. However, note that it is not always necessary to segment an image in order to extract summary information from it. For example, areas of different tissue types could have been estimated directly from the bivariate histogram of pixel values in the MRI image. Similarly, areas of different land use can be estimated from the distribution of pixel values in the multivariate Landsat image. We shall return to this example in §6.1.

4.1.3 Contextual classifiers

Thresholding is most successful when there is little overlap in distributions of pixel values from the different categories in an image. Noise is one cause of overlap, which can be reduced by using a smoothing filter. Unfortunately, filters—linear ones in particular—also blur edges to some extent, and therefore produce pixels with values intermediate between category means. To illustrate the effects of filtering, Fig. 4.7(a) shows the histogram of the muscle fibres image (Fig. 1.6b), and Fig. 4.7(b) shows the histogram after the application of a Gaussian filter with $\sigma^2 = 6\frac{2}{3}$ (§3.1.1) to the image. Both histograms are displayed on square-root scales in order to reveal the details in small values. Note that Fig. 4.7(b) shows a greater number of pixels with values intermediate between the two peaks. Usually, smoothing would also produce narrower peaks in the histogram, but this is not the case here because noise levels are low.

Kirby and Rosenfeld (1979) proposed forming a bivariate histogram by plotting pixel values in the original image against those from its filtered form. For the muscle fibres image, this is shown in Fig. 4.7(c). (The histograms in Figs. 4.7(a) and (b) are the marginal histograms obtainable by summing over columns or rows in Fig. 4.7(c).) Some denser regions of pixel values are visible on the diagonal joining the two main clusters of points.

If a histogram is formed only from pixels within 5 of the diagonal of this plot then Fig. 4.7(d) results. The histogram has been sharpened in comparison with Figs 4.7(a, b), to the extent that additional, small peaks have become visible. Figure 4.8 shows the effect of multiple thresholding of the muscle fibres image, at values of 50, 90, 130 and 165, which are indicated by arrows in Fig. 4.7(d). Pixels outside the range 50 – 165 are displayed as white. Within this range, pixels are displayed as three shades of grey of increasing lightness in the three intervals 51–90, 91–130 and 131–165. The subpeaks can be seen to be from pixels on the boundaries between fibres and from four muscle fibres with pixel values intermediate between the two main fibre types. These are the third type of fibre, the fast-twitch glycolytic ones mentioned in §1.1.3.

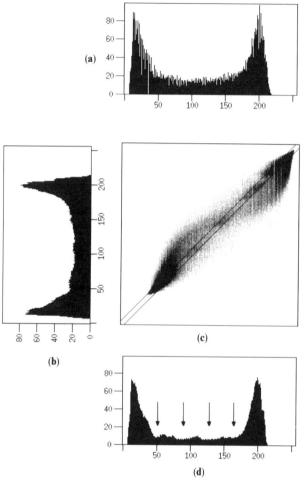

Fig. 4.7 Histograms of muscle fibres image: (a) original pixel values, displayed on a square-root scale; (b) image after smoothing by Gaussian filter ($\sigma^2 = 6\frac{2}{3}$); (c) bivariate histogram of original pixel values and those after smoothing, with all counts exceeding 25 displayed as black and the parallel diagonal lines indicating the points extracted for (d); (d) histogram of pixel values after smoothing those values within 5 of those before smoothing (the arrows indicate the thresholds used in Fig. 4.8).

Rosenfeld and co-workers published several variants on the basic idea of a bivariate histogram. The work culminated in the paper by Weszka and Rosenfeld (1979), in which it was advocated that the bivariate histogram should be formed from image values and output from an edge filter. This can be viewed as a variant of Fig. 4.7(c), because the difference between an image and its smoothed form is a type of edge filter. The general principle is that, by constructing a histogram from non-edge pixels only its peaks should be clearer.

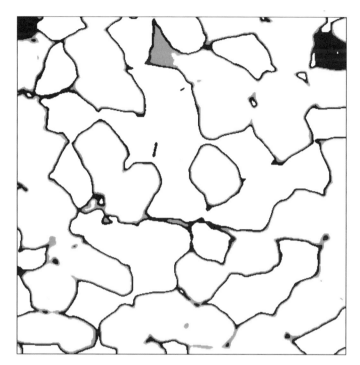

Fig. 4.8   Thresholded muscle fibres image. Pixels in the ranges 0 – 50 and 166 – 255 are displayed as white, those between 51 and 90 are black, those between 91 and 130 are dark grey, and those between 131 and 165 are light grey.

Post-classification smoothing

Another way of using contextual information is by applying a *majority filter* to a segmented image, termed *post-classification smoothing*. For example, if pixel (i,j) has been labelled as category g_{ij} by a segmentation algorithm then a majority filter relabels (i,j) as the most common category in a $(2m + 1) \times (2m + 1)$ window centred on pixel (i,j), i.e. the set $\{g_{i+k, j+l}$ for $k,l = -m, \ldots, m\}$.

Contextual information can also be used in automatic segmentation algorithms. Mardia and Hainsworth (1988) investigated incorporating post-classification smoothing into a thresholding algorithm. (Their paper also provides a succinct review of algorithms for thresholding images whose pixel values are mixtures of Gaussian distributions.) To explain the method, we shall consider the simplest case. First we shall discuss it in words, and then more formally.

From an initial guess at a threshold, an image is segmented. Then the classified image is smoothed using a majority filter, so that neighbouring pixels are more often classified the same. Mean pixel values are evaluated in each category, and the threshold is re-estimated using the intermeans criterion (§4.1.1). The image is segmented and smoothed again, and so on until the threshold does not change between successive iterations. Formally, we have the following:

1. Make an initial guess at a threshold t.
2. Segment the image, by thresholding, and store the result in an array g:

$$g_{ij} = 1 \quad \text{if } f_{ij} \leq t, \quad \text{otherwise } g_{ij} = 2.$$

3. Apply a majority filter to g, and record the new labels in array g'. Therefore g'_{ij} is set to the most common category in the set $\{g_{i+k,j+l}$ for $k, l = -m, \ldots, m\}$. This is repeated for every value of i and j from $m + 1$ to $n - m$. (Ranges have been chosen to avoid image borders, as in Chapter 3.)

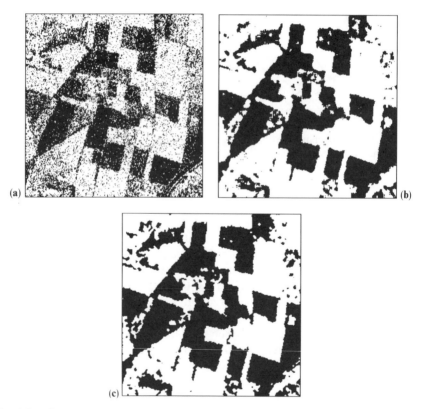

Fig. 4.9   Three automatic segmentations of the log-transformed SAR image: (a) obtained by thresholding using the intermeans algorithm; (b) thresholding combined with a post-classification smoothing by majority filter in a 5 × 5 window; (c) iterated conditioned modes (ICM) classification, with $\beta = 1.5$.

4. Calculate the mean pixel value in category 1,

$$\mu_1 = \frac{1}{N_1} \sum\sum_{\{(i,j) \text{ such that } g'_{ij}=1\}} f_{ij},$$

   where N_1 is the number of pixels in category 1. Similarly, calculate μ_2.

5. Re-estimate t as $[\frac{1}{2}(\mu_1 + \mu_2)]$.

6. Repeat steps 2–5 until t converges to a stable value, and then stop after step 3.

Figure 4.9(b) shows the segmentation of the log-transformed SAR image (as shown in Fig. 2.6) into two classes using the above algorithm and a 5×5 smoothing window. The algorithm converged in four iterations, starting with a threshold set at the median pixel value. The window size was chosen by trial and error: a size less than 5×5 was inadequate for smoothing the inconsistencies in classification caused by the noise, whereas a larger size produced an over-smooth segmentation—corners of fields became rounded. Figure 4.9(a) shows the result of the same algorithm without smoothing— that is, the intermeans algorithm. The noise is so great in this image that the histogram of pixel values is unimodal, and no threshold can achieve a better segmentation.

ICM algorithm

Mardia and Hainsworth compared their algorithm with a simple form of *Bayesian image restoration*, which uses Besag's (1986) *iterated conditional modes (ICM)* algorithm. We shall consider the simplest case of ICM, first in words and then formally.

From an initial guess at a threshold, or equivalently of the two category means, the image is segmented. Mean pixel values are evaluated in each category, and the image is re-segmented, pixel-by-pixel, using a variable threshold that takes account of both the category means and the current classification of neighbouring pixels. Category means are recalculated and the recursive segmentation procedure is re-applied. These steps are repeated until the segmentation remains (almost) unchanged on successive iterations. More formally, we have the following:

1. Make an initial guess at μ_1 and μ_2, and set

$$g_{ij} = 1 \quad \text{if } (f_{ij} - \mu_1)^2 \leq (f_{ij} - \mu_2)^2, \qquad \text{otherwise } g_{ij} = 2.$$

   Or, equivalently,

$$g_{ij} = 1 \quad \text{if } f_{ij} \leq t, \qquad \text{where } t = \left[\frac{\mu_1 + \mu_2}{2}\right].$$

2. Calculate the mean pixel values in the two categories, μ_1 and μ_2. Also calculate the within-category variance

$$\sigma^2 = \frac{1}{n^2} \sum_{i=1}^{n} \sum_{j=1}^{n} (f_{ij} - \mu_{g_{ij}})^2.$$

3. Re-segment, recursively, as follows. For each value of i between 1 and n, apply the following criterion in turn for j between 1 and n:

$$g_{ij} = 1 \quad \text{if} \quad (f_{ij} - \mu_1)^2 - \beta\sigma^2 N_{ij} \leq (f_{ij} - \mu_2)^2 - \beta\sigma^2 (8 - N_{ij}),$$

where N_{ij} is the number of the eight neighbours of pixel (i, j), currently classified as 1, and β is a constant whose value must be chosen. A value of 1.5 has often been used in the literature.

Equivalently, this reclassification be viewed as a variable threshold:

$$g_{ij} = 1 \quad \text{if } f_{ij} \leq t_{ij}, \quad \text{where } t_{ij} = \left[\frac{(\mu_1 + \mu_2)}{2} + \frac{\beta\sigma^2 (2N_{ij} - 8)}{2(\mu_2 - \mu_1)} \right].$$

Note that $(i-1, j-1), (i-1, j), (i-1, j+1)$ and $(i, j-1)$ will have already been updated, although, alternatively, all pixels can be reclassified simultaneously. The two approaches are termed *asynchronous* and *synchronous*, respectively.

4. Repeat steps 2 and 3 until no (or only a few) pixels change category.

Figure 4.9(c) shows the results of applying the ICM algorithm to the SAR image. In this case, 20 iterations were performed, by which stage fewer than 5 pixels changed classes between iterations. The final segmentation is very similar to that produced by applying the majority filter, shown in Fig. 4.9(b). Mardia and Hainsworth (1988) also found that the two algorithms often gave very similar results.

The ICM algorithm is a very simple example of Bayesian image restoration—a potentially powerful technique that has received much attention from statisticians in the last decade, following seminal papers by Geman and Geman (1984) and Besag (1986). The underlying principle is that of combining

> *prior information* on what the segmented image should be like, and
> *data*, the observed pixel values in an image,

in order to produce a segmentation. In our example, the term βN_{ij} is motivated by a *Markov random field model*, and favours a segmentation in which neighbouring pixels are classified the same. The main attraction of Bayesian image restoration is that it is open to many refinements, such as incorporating more specific prior information. However, computational feasibility of Bayesian restoration algorithms can be a limiting factor. See e.g., Aykroyd and Green

(1991), Ripley and Sutherland (1990) and Qian and Titterington (1991) for a range of applications.

4.2   EDGE-BASED SEGMENTATION

As we have seen, the results of threshold-based segmentation are usually less than perfect. Often, a scientist will have to make changes to the results of automatic segmentation. One simple way of doing this is by using a computer mouse to control a screen cursor and draw boundary lines between regions. Figure 4.10(a) shows the boundaries obtained by thresholding the muscle fibres image (as already displayed in Fig. 4.1a), superimposed on the output from Prewitt's edge filter (§3.4.2), with the contrast stretched so that values between 0 and 5 are displayed as shades of grey ranging from white to black and values exceeding 5 are all displayed as black. This display can be used as an aid to determine where extra boundaries need to be inserted to fully segment all muscle fibres. Figure 4.10(b) shows the result after manually adding 71 straight lines.

Algorithms are available for semi-automatically drawing edges, whereby the scientist's rough lines are smoothed and perturbed to maximize some

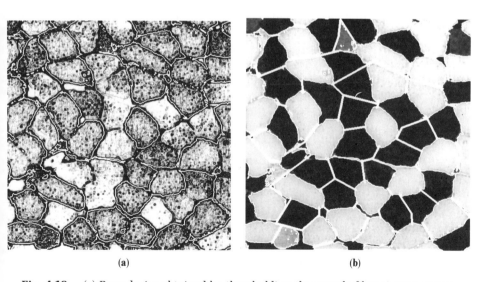

(a)   (b)

Fig. 4.10   (a) Boundaries obtained by thresholding the muscle fibres image, superimposed on output for Prewitt's filter, with values between 0 and 5 displayed in progressively darker shades of grey, and values in excess of 5 displayed as black. (b) Manual segmentation of the image by addition of extra lines to boundaries obtained by thresholding, superimposed in white on the original image.

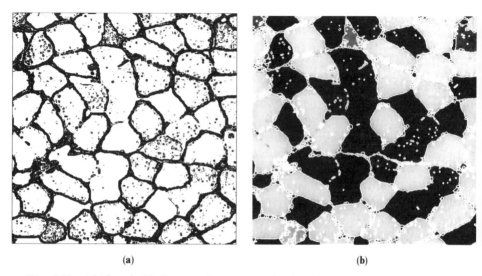

(a)   (b)

Fig. 4.11   (a) Thresholded output from Prewitt's edge filter applied to muscle fibres image: values greater than 5 are displayed as black, those less than or equal to 5 as white. (b) Boundaries produced from connected regions in (a), superimposed in white on the original image.

criterion of match with the image (e.g., Samadani and Han, 1993). Alternatively, edge finding can be made fully automatic, although not necessarily fully successful. Figure 4.11(a) shows the result of applying Prewitt's edge filter to the muscle fibre image. In this display, the filter output has been thresholded at a value of 5: all pixels exceeding 5 are labelled as edge pixels and displayed as black. Connected chains of edge pixels divide the image into regions. Segmentation can be achieved by allocating to a single category all non-edge pixels that are not separated by an edge. Rosenfeld and Pfaltz (1966) gave an efficient algorithm for doing this for 4- and 8-connected regions, termed a *connected components algorithm*. We shall describe this algorithm in words, and then mathematically.

Connected components algorithm

The algorithm operates on a *raster scan*, in which each pixel is visited in turn, starting at the top-left corner of the image and scanning along each row, finishing at the bottom-right corner. For each non-edge pixel, (i, j), the following conditions are checked. If its already-visited neighbours---$(i - 1, j)$ and $(i, j - 1)$ in the 4-connected case, also $(i - 1, j - 1)$ and $(i - 1, j + 1)$ in the 8-connected case --- are all edge pixels then a new category is created and (i, j) is allocated to it. Alternatively, if all its non-edge neighbours are in a single category then (i, j) is also placed in that category. The final possibility is that neighbours belong to two or more categories, in which case (i, j) is allocated to one of them and a note is kept that these categories are connected and therefore from then on should be considered as a single category. More formally, for the simpler case of 4-connected regions, we have the following:

- Initialize the count of the number of categories by setting $K = 0$.
- Consider each pixel (i, j) in turn in a raster scan, proceeding row-by-row $(i = 1, \ldots, n)$, and for each value of i taking $j = 1, \ldots, n$.
- One of four possibilities applies to pixel (i, j):
  1. If (i, j) is an edge pixel then nothing needs to be done.
  2. If both previously visited neighbours, $(i - 1, j)$ and $(i, j - 1)$, are edge pixels then a new category has to be created for (i, j):

$$K \rightarrow K + 1, \quad h_K = K, \quad g_{ij} = K,$$

  where the entries in $h_1, \ldots, h_K$ are used to keep track of which categories are equivalent, and g_{ij} records the category label for pixel (i, j).
  3. If just one of the two neighbours is an edge pixel then (i, j) is assigned the same label as the other one:

$$g_{ij} = \begin{cases} g_{i-1, j} & \text{if } (i, j - 1) \text{ is the edge pixel,} \\ g_{i, j-1} & \text{otherwise.} \end{cases}$$

  4. The final possibility is that neither neighbour is an edge pixel, in which case (i, j) is given the same label as one of them,

$$g_{ij} = g_{i-1, j},$$

  and if the neighbours have labels that have not been marked as equivalent, i.e. $h_{g_{i-1, j}} \neq h_{g_{i, j-1}}$, then this needs to be done (because they are connected at pixel (i, j)). The equivalence is recorded by changing the entries in $h_1, \ldots, h_K$, as follows:
  — set $l_1 = \min \left(h_{g_{i-1, j}}, h_{g_{i, j-1}} \right)$ and $l_2 = \max \left(h_{g_{i-1, j}}, h_{g_{i, j-1}} \right)$;
  — for each value of k from 1 to K, if $h_k = l_2$ then $h_k \rightarrow l_1$.
- Finally, after all the pixels have been considered, the array of labels is revised, taking into account which categories have been marked for amalgamation:

$$g_{ij} \rightarrow h_{g_{ij}} \quad \text{for } i, j = 1, \ldots, n.$$

After application of the labelling algorithm, superfluous edge pixels—that is, those that do not separate classes—can be removed: any edge pixel that has neighbours only of one category is assigned to that category.

Figure 4.11(b) shows the result of applying the labelling algorithm with edges as shown in Fig. 4.11(a), and removing superfluous edge pixels.

The white boundaries have been superimposed on the original image. Similarly, small segments (say less than 500 pixels in size) that do not touch the borders of the image can be removed, leading to the previously displayed Fig. 4.1(b). The segmentation has done better than simple thresholding, but has failed to separate all fibres because of gaps in output from Prewitt's edge filter. Martelli (1976), among others, has proposed algorithms for bridging these gaps.

Laplacian-of-Gaussian filter

Another edge-detection filter considered in Chapter 3 was the Laplacian-of-Gaussian (§3.1.2). This filter can also be used to segment images, using the zero-crossings of the output to specify positions of boundaries. One advantage over Prewitt's filter is that the zero-crossings always form closed boundaries. Figure 4.12(a) shows output from the Laplacian-of-Gaussian filter ($\sigma^2 = 27$), applied to the muscle fibres image. Boundaries corresponding to weak edges can be suppressed by applying a threshold to the average gradient strength around a boundary. Figure 4.12(b) shows zero-crossings of the Laplacian-of-Gaussian filter with average gradients exceeding unity, superimposed on the original image. In this application, the result can be seen to be little better than simple thresholding.

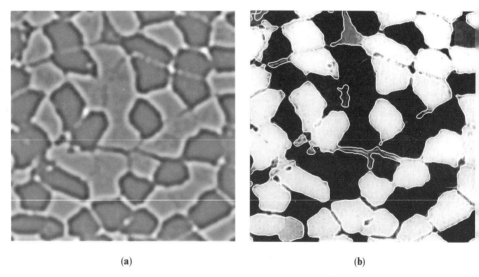

(a)   (b)

Fig. 4.12   (a) Output from Laplacian-of-Gaussian filter ($\sigma^2 = 27$) applied to muscle fibres image. (b) Zero-crossings from (a) with average image gradients in excess of 1.0, in white superimposed on the image.

4.3   REGION-BASED SEGMENTATION

Segmentation may be regarded as *spatial clustering*:

- *clustering* in the sense that pixels with similar values are grouped together; and
- *spatial* in that pixels in the same category also form a single connected component.

Clustering algorithms may be agglomerative, divisive or iterative (see e.g., Gordon, 1981). Region-based methods can be similarly categorized into

- those that *merge* pixels;
- those that *split* the image into regions; and
- those that both *split and merge* in an iterative search scheme.

The distinction between edge-based and region-based methods is a little arbitrary. For example, in §4.2, one of the algorithms we considered involved placing all neighbouring non-edge pixels in the same category. In essence, this is a merging algorithm.

Watershed algorithm

Seeded region growing is a semi-automatic method of the merge type. We shall explain it by way of an example. Figure 4.13(a) shows a set of *seeds*, white discs of radius 3, which have been placed inside all the muscle fibres, using an on-screen cursor controlled by a computer mouse. Figure 4.13(b) shows again the output from Prewitt's edge filter. Superimposed on it in white are the seeds and the boundaries of a segmentation produced by a form of *watershed algorithm*. The boundaries are also shown superimposed on the original muscle fibres image in Fig. 4.13(c). The watershed algorithm operates as follows (we shall explain the name later).

For each of a sequence of increasing values of a threshold, all pixels with edge strength less than this threshold that form a connected region with one of the seeds are allocated to the corresponding fibre. When a threshold is reached for which two seeds become connected, the pixels are used to label the boundary. A mathematical representation of the algorithm is too complicated to be given here. Instead, we refer the reader to Vincent and Soille (1991) for more details and an efficient algorithm. Meyer and Beucher (1990) also consider the watershed algorithm, and added some refinements to the method.

Note the following:

- The use of discs of radius 3 pixels, rather than single points, as seeds makes the watershed results less sensitive to fluctuations in Prewitt's filter output in the middle of fibres.

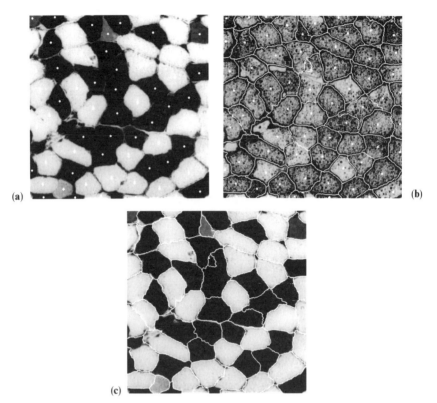

(a)

(b)

(c)

Fig. 4.13   Manual segmentation of muscle fibres image by use of watershed algorithm: (a) manually positioned 'seeds' in centres of all fibres; (b) output from Prewitt's edge filter, together with watershed boundaries; (c) watershed boundaries superimposed on the image.

- The results produced by this semi-automatic segmentation algorithm are almost as good as those shown in Fig. 4.10(b), but the effort required in positioning seeds inside muscle fibres is far less than that required to draw boundaries.
- Adams and Bischof (1994) present a similar seeded-region growing algorithm, but based directly on the image greyscale, not on the output from an edge filter.

The watershed algorithm, in its standard use, is fully automatic. Again, we shall demonstrate this by illustration. Figure 4.14 shows the output produced by a variance filter (§3.4.1) with Gaussian weights ($\sigma^2 = 96$) applied to the muscle fibres image after histogram equalization (as shown in Fig. 2.7d). The white seeds overlie all the local minima of the filter output, that is, pixels whose neighbours all have larger values and so are shaded lighter. Note that it is necessary to use a large value of σ^2 to ensure that the filter

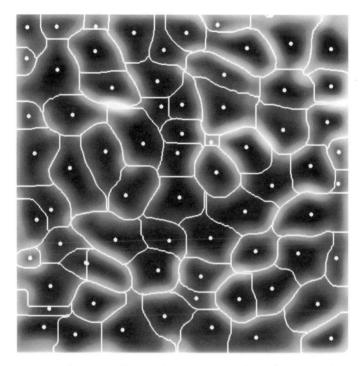

Fig. 4.14   Output of variance filter with Gaussian weights ($\sigma^2 = 96$) applied to muscle fibres image, together with seeds indicating all local minima and boundaries produced by watershed algorithm.

output does not have many more local minima. The boundaries produced by the watershed algorithm have been added to Fig. 4.14. An intuitive way of viewing the watershed algorithm is by considering the output from the variance filter as an *elevation map*: light areas are high ridges and dark areas are valleys (as in Fig. 1.5). Each local minimum can be thought of as the point to which any water falling on the region drains, and the segments are the catchments for them. Hence, the boundaries, or watersheds, lie along tops of ridges. Figure 4.1(c) shows this segmentation superimposed on the original image.

Split-and-merge algorithm

There are very many other region-based algorithms, but most of them are quite complicated. In this section, we shall consider just one more, namely an elegant split-and-merge algorithm proposed by Horowitz and Pavlidis (1976). We shall present it in a slightly modified form to segment the log-transformed SAR image (Fig. 2.6), basing our segmentation decisions on variances,

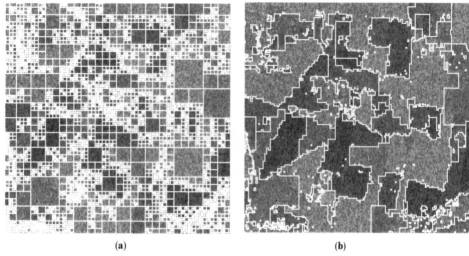

(a)   (b)

Fig. 4.15   Region-growing segmentation of log-transformed SAR image: (a) division of image into squares with variance less than 0.60, obtained as first step in algorithm; (b) final segmentation, after amalgamation of squares, subject to variance limit of 0.60.

whereas Horowitz and Pavlidis based theirs on the range of pixel values. The algorithm operates in two stages, and requires a limit to be specified for the maximum variance in pixel values in a region.

The first stage is the *splitting* one. Initially, the variance of the whole image is calculated. If this variance exceeds the specified limit then the image is subdivided into four quadrants. Similarly, if the variance in any of these four quadrants exceeds the limit, it is further subdivided into four. This continues until the whole image consists of a set of squares of varying sizes, all of which have variances below the limit. (Note that the algorithm must be capable of achieving this because at the finest resolution of each square consisting of a single pixel the variances are taken to be zero.)

Figure 4.15(a) shows the resulting boundaries in white, superimposed on the log-transformed SAR image, with the variance limit set at 0.60. Note the following:

- Squares are smaller in non-uniform parts of the image.
- The variance limit was set to 0.60, rather than to the speckle variance value of 0.41 (Horgan, 1994), because in the latter case the resulting regions were very small.
- The algorithm requires the image dimension n to be a power of 2. Therefore the 250×250 SAR image was filled out to 256×256 by adding borders of width 3.

The second stage of the algorithm, the *merging* one, involves amalgamating squares that have a common edge, provided that by so doing the variance of the new region does not exceed the limit. Once all amalgamations have been completed, the result is a segmentation in which every region has a variance less than the set limit. However, although the result of the first stage in the algorithm is unique, that from the second is not—it depends on the order in which squares are considered.

Figure 4.15(b) shows the boundaries produced by the algorithm, superimposed on the SAR image. Dark and light fields appear to have been successfully distinguished, although the boundaries are rough and retain some of the artefacts of the squares in Fig. 4.15(a).

Pavlidis and Liow (1990) proposed overcoming the deficiencies in the boundaries produced by the Horowitz and Pavlidis algorithm by combining the results with those from an edge-based segmentation. Many other ideas for region-based segmentation have been proposed (e.g., the review of Haralick and Shapiro, 1985), and it is still an active area of research.

One possibility for improving segmentation results is to use an algorithm that over-segments an image, and then apply a rule for amalgamating these regions. This requires 'high-level' knowledge, which falls into the domain of artificial intelligence. (All that we have considered in this chapter may be termed 'low-level'.) For applications of these ideas in the area of remote sensing, see Tailor *et al.* (1986) and Ton, Sticklen and Jain (1991). It is possible that such domain-specific knowledge could be used to improve the automatic segmentations of the SAR and muscle fibres images, for example by constraining boundaries to be straight in the SAR image and by looking only for convex regions of specified size for the muscle fibres.

We briefly mention some other, more-complex techniques that can be used to segment images.

- The *Hough transform* (see e.g., Leavers, 1992) is a powerful technique for finding straight lines, and other parametrized shapes, in images.
- Boundaries can be constrained to be smooth by employing roughness penalties such as bending energies. The approach of varying a boundary until some such criterion is optimized is known as the fitting of *snakes* (Kass, Witkin and Terzopoulos 1988).
- Models of expected shapes can be represented as *templates* and matched to images. Either the templates can be rigid and the mapping can be flexible (e.g. the thin-plate spline of Bookstein, 1989), or the template itself can be flexible, as in the approach of Amit, Grenander and Piccioni (1991).
- Images can be broken down into fundamental shapes, in a way analogous to the decomposition of a sentence into individual words, using *syntactic* methods (Fu, 1974).

4.4 SUMMARY

The key points about image segmentation are as follows:

- Segmentation is the allocation of every pixel in an image to one of a number of categories, which correspond to objects or parts of objects. Commonly, pixels in a single category should
  — have similar pixel values;
  — form a connected region in the image;
  — be dissimilar to neighbouring pixels in other categories.
- Segmentation algorithms may either be applied to the original images, or after the application of transformations and filters (considered in Chapters 2 and 3).
- Three general approaches to segmentation are
  — thresholding;
  — edge-based methods;
  — region-based methods.
- Methods within each approach may be further divided into those that:
  — require manual intervention, or
  — are fully automatic.

A single threshold t operates by allocating pixel (i, j) to category 1 if $f_{ij} \leq t$, and otherwise putting it in category 2. Thresholds may be obtained by

- manual choice, or
- applying an algorithm such as intermeans or minimum-error to the histogram of pixel values:
  — intermeans positions t half-way between the means in the two categories;
  — minimum-error chooses t to minimize the total number of misclassifications on the assumption that pixel values in each category are normally distributed.

- Thresholding methods may also be applied to multivariate images. In this case, two possibilities are
  — manually selecting a training set of pixels that are representative of the different categories, and then using linear discrimination;
  — k-means clustering, in which the categories are selected automatically from the data.
- The context of a pixel, that is the values of neighbouring pixels, may also be used to modify the threshold value in the classification process. We considered three methods:
  — restricting the histogram to those pixels that have similarly valued neighbours;
  — post-classification smoothing;
  — using Bayesian image restoration methods, such as the iterated conditional modes (ICM) algorithm.

In edge-based segmentation, all pixels are initially labelled as either being on an edge or not; then non-edge pixels that form connected regions are allocated to the same category. Edge labelling may be

- manual, by using a computer mouse to control a screen cursor and draw boundary lines between regions;
- automatic, by using an edge-detection filter; edges can be located either
  — where output from a filter such as Prewitt's exceeds a threshold, or
  — at zero crossings from a Laplacian-of-Gaussian filter.

Region-based algorithms act by grouping neighbouring pixels that have similar values, and splitting groups of pixels that are heterogeneous in value. Three methods were considered:

- Regions may be grown from manually positioned 'seed' points, for example by applying a watershed algorithm to output from Prewitt's filter.
- The watershed algorithm may also be run fully automatically, for example by using local minima from a variance filter as seed points.
- One split-and-merge algorithm finds a partition of an image such that the variance in pixel values within every segment is below a specified threshold, but no two adjacent segments can be amalgamated without violating the threshold.

Finally, the results from automatic segmentation can be improved by

- using methods of mathematical morphology (Chapter 5);
- using domain-specific knowledge, which is beyond the scope of this book

The segmentation results will be used in Chapter 6 to extract quantitative information from images.

5

Mathematical Morphology

Mathematical morphology is a theory that provides a number of useful tools for image analysis. It is seen by some as a self-contained approach to handling images and by others (including the authors) as complementary to the other methods presented in this book. Figure 5.1 shows an example of an operation based on mathematical morphology (hereinafter referred to simply as morphology). Figure 5.1(a) shows a subset of the soil image thresholded at a pixel value of 20 (Fig. 4.2d). Figure 5.1(b) is the result of a morphological operation termed a *closing*, which is described in §5.1. At first, it may seem to be just another filter that has smoothed the image, and in fact the closing operation is often used for this purpose. However, it does have different properties from the sort of filters considered in Chapter 3. For example, although many black pixels have been changed to white to smooth the image, none have been changed from white to black.

Morphology is an approach to image analysis that is based on the assumption that an image consists of structures that may be handled by set theory. This is unlike most of the rest of the ideas in this book, which are based on arithmetic. It is based on ideas developed by J. Serra and G. Matheron of the Ecole des Mines in Fontainebleau, France. The seminal work is Serra (1982). Other useful introductions may be found in Serra (1986), Haralick, Sternberg and Zhuang (1987) and Haralick and Shapiro (1992, Ch. 5). Morphology has become popular in recent years. The basic operations are available in many image analysis software packages. Morphology lends itself to efficient parallel hardware implementations, and computers using this are available.

This chapter develops the ideas of morphology mainly for binary images. Many applications use thresholded versions of greyscale images, usually where noise is not a dominant feature. The extension to greyscale images will also be covered. This will be done by regarding a greyscale image as a binary image in three dimensions (§5.5).

As morphological operations are based on sets, set notation will be used in this chapter. A reminder of the basic concepts is given in Fig. 5.2, using sets of lattice points. Sets are simply groups of pixels, and the terminology is just a

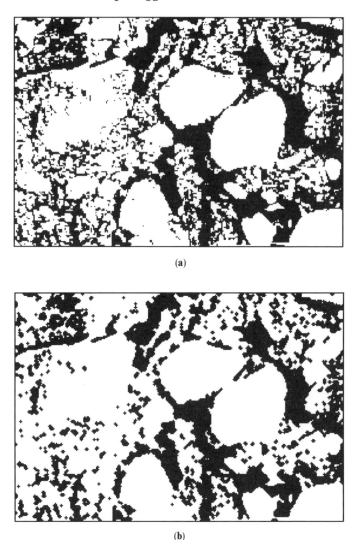

(a)

(b)

Fig. 5.1   (a) Subset of the soil image thresholded at 120. (b) Morphological closing of (a) using a lattice approximation to a disc of radius 1

convenient way of describing what pixels lie in particular groups. (The reader interested in acquiring a fuller knowledge of set theory should read Vilenkin (1968).) However, all definitions in set notation will also be explained in words. Section 5.1 introduces the basic ideas of morphology, §5.2–5.4 look at operations which affect particular aspects of images, and §5.5 describes how the ideas may be extended to greyscale images. In §5.6, we shall summarize the key results.

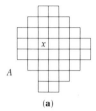

(a)

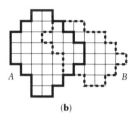

(b)

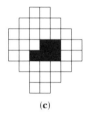

(c)

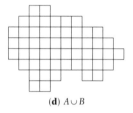

(d) $A \cup B$

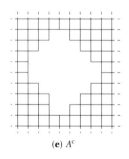

(e) A^c

(f) $A \cap B$

Fig. 5.2   Basic ideas of set theory. (a) The set A. The pixel x is an element of A (written $x \in A$). (b) Two overlapping sets A and B. (c) Subsets: the shaded pixels C are a subset of A (written $C \subset A$). (d) The union of A and B (written $A \cup B$). (e) The complement of A (written A^c). (f) The intersection of A and B (written $A \cap B$).

5.1 BASIC IDEAS

Morphology is based on set theory, and so the fundamental objects are sets. For morphology of binary images, the sets consist of pixels in an image. Readers interested in the full mathematical background are referred to Serra (1988).

Figure 5.3(a) shows an example of an image containing three sets of black pixels. Either the pixels labelled 0 (displayed as black) or those labelled 1 (displayed as white) in a binary image may comprise the sets of interest. Unless

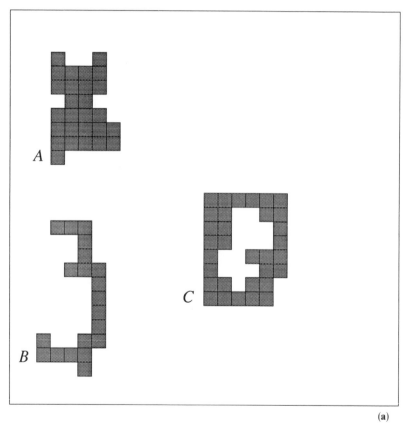

(a)

 D

(b)

Fig. 5.3   (a) Example of some sets with different shapes. (b) A test set. The • in the top left pixel indicates that we have selected this as the reference pixel

stated otherwise, we shall take the black pixels to be the sets of interest. In this chapter, when we refer to operations on the *image*, we shall be referring to operations on the set of all black pixels. Usually this will be the union of several separate sets of black pixels—what we think of as individual *objects*, such as *A*, *B* and *C* in Fig. 5.3(a). The white pixels are the *complement* (Fig. 5.2e) of the black pixels, the complement of a set being the set of elements it does not contain. Any operation that affects the set of black pixels will also affect the set of white pixels. For example, removing a pixel from the set of black pixels naturally creates a new white pixel.

We can see that the sets in Fig. 5.3(a) have different shapes. For example, set *B* is about the same size (contains about as many pixels) as set *A*, but is a different shape—it is longer and thinner. One way of describing such differences is in terms of intersections with 'test' sets. If we let *D* be the simple 2×2 set shown in Fig. 5.3(b) then it is possible to *translate D* to positions such that $D \subset A$, whereas no translations can be found to satisfy $D \subset B$. (Here $\subset$ denotes 'is a subset of' : see Fig. 5.2(c).) This is a consequence only of the shapes of *A* and *B*, and so such tests provide a means of analysing shape information in the image. Test sets such as *D* will be termed *structuring elements*. It is an important idea that these structuring elements can be placed at any pixel in an image (although rotation of the structuring element is not allowed.) To do this, we use some *reference pixel* whose position defines where the structuring element has been placed. The choice of this reference pixel is often arbitrary, it need not even lie in *D*. We may choose, for example, the centre pixel if the set is symmetric, or one of the corners if it is a polygon. (The only effect of how this choice is made is to translate the position of the result of a morphological operation.)

Morphological operations transform the image. The most basic morphological operation is that of *erosion*. Suppose *A* is a set (a binary image or part of it) and *S* is a structuring element. If *S* is placed with its reference pixel at (i, j), we denote it by $S_{(i, j)}$. Then the erosion of *A* by *S* is defined to be the set of all pixel locations for which *S* placed at that pixel is contained within *A*. This is denoted $A \ominus S$, and may be written

$$A \ominus S = \{(i, j) : S_{(i, j)} \subset A\}.$$

For the example in Fig. 5.3, let $S = D$, a block of four pixels. If we use the top-left corner pixel of D as the reference pixel then the result of the erosion $I \ominus D$, where *I* denotes the whole image, i.e. $A \cup B \cup C$, is as shown in Fig. 5.4. To see how the erosion may be performed on an image, we first note that if *D* is placed with its reference pixel at (i, j), then $D_{(i, j)}$ consists of the four pixels $(i, j), (i + 1, j), (i, j + 1)$ and $(i + 1, j + 1)$, i.e.

$$D_{(i, j)} = \{(i, j), (i + 1, j), (i, j + 1), (i + 1, j + 1)\}.$$

If (and only if) all of these are black will the pixel (i, j) in the eroded image be black. If we let *g* be the eroded image then an algorithm to perform the erosion

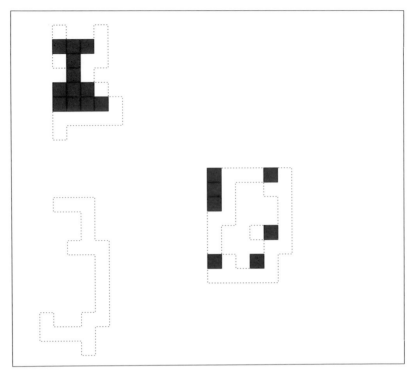

Fig. 5.4   The image in Fig. 5.3(a) after erosion with the structuring element D.

will operate as follows, recalling that $f_{ij} = 0$ or 1 according to whether the pixel is black or white:

1. Set $g_{ij} = f_{ij}$. If $f_{ij} = 1$, go to step 5.
2. If $f_{i+1,j} = 1$, set $g_{ij} = 1$ and go to step 5.
3. If $f_{i,j+1} = 1$, set $g_{ij} = 1$ and go to step 5.
4. If $f_{i+1,j+1} = 1$, set $g_{ij} = 1$ and go to step 5.
5. Move to the next pixel in the image and go to step 1.

The above algorithm is efficient and easy to generalise to erosions with other sets. A simpler to program, but less efficient, algorithm for this erosion would be to compute

$$g_{ij} = 1 - (1 - f_{ij})(1 - f_{i+1,j})(1 - f_{i,j+1})(1 - f_{i+1,j+1}),$$

since g_{ij} will be zero only if $f_{ij}, f_{i+1,j}, f_{i,j+1}$ and $f_{i+1,j+1}$ are all zero. This is less efficient, since it involves arithmetic, which is usually slower on a computer than making comparisons.

A complementary operation to that of erosion is *dilation*. It is defined simply as the erosion of the complement of a set. If A^c denotes the complement of A

then the dilation of a set A by a set S, denoted by $A \oplus S$, is defined by

$$A \oplus S = (A^c \ominus S)^c.$$

This definition easily leads to algorithms for obtaining the dilation.

There is an alternative, equivalent definition of the dilation. Let $S'_{(i,j)}$ denote the reflection of $S_{(i,j)}$ in (i, j), i.e. the rotation of S through $180°$ about (i, j). Then $A \oplus S$ contains all the pixels lying in any $S'_{(i,j)}$ for which $(i, j) \in A$. We can think of the dilation as placing a copy of S' at every pixel in A. If the reference pixel is symmetrically placed then $S' = S$. We may write this definition as:

$$A \oplus S = \bigcup_{(i,j)\in A} S'_{(i,j)}$$

where $\cup$ denotes set union (Fig. 5.2d).

Erosion and dilation are illustrated in Figs 5.5(c) and (d). The binary image used here is a subset of the the turbinate image (Fig. 1.2a).

Two widely used operations are those of *opening* and *closing*, which are often denoted by $\psi_S(A)$ and $\phi_S(A)$ respectively. These are defined as

$$\psi_S(A) = (A \ominus S) \oplus S',$$
$$\phi_S(A) = (A \oplus S) \ominus S',$$

so that an opening is an erosion *followed* by a dilation, and a closing is a dilation followed by an erosion. They are complementary, in that applying one to A is equivalent to applying the other to A^c. Another equivalent definition of the opening is that it is the union of all sets $S_{(i,j)}$ that are contained in A:

$$\psi_S(A) = \bigcup_{S_{(i,j)}\subset A} S_{(i,j)}.$$

This differs from the erosion, which consists of only the reference pixels, rather than the whole set λ, for which $S_{(i,j)}$ is contained in A. The closing is defined similarly, but applied to A^c. These operations are illustrated in Fig. 5.5. For the structuring element used here, both have the effect of smoothing the image. The opening does this by removing pixels from the set, the closing by adding them. Note that both opening and closing are *idempotent* operations; that is, applying them more once produces no further effect:

$$\psi_S(\psi_S(A)) = \psi_S(A)$$

and

$$\phi_S(\phi_S(A)) = \phi_S(A).$$

A more general morphological operation than erosion is the *hit-or-miss transform*. The structuring element is a set with two components, $S^1_{(i,j)}$ and $S^2_{(i,j)}$, placed so that both reference pixels are at position (i, j). The hit-or-miss transform of a set A is then defined as

$$A \circledast S = \{(i, j) : S^1_{(i,j)} \subset A; S^2_{(i,j)} \subset A^c\};$$

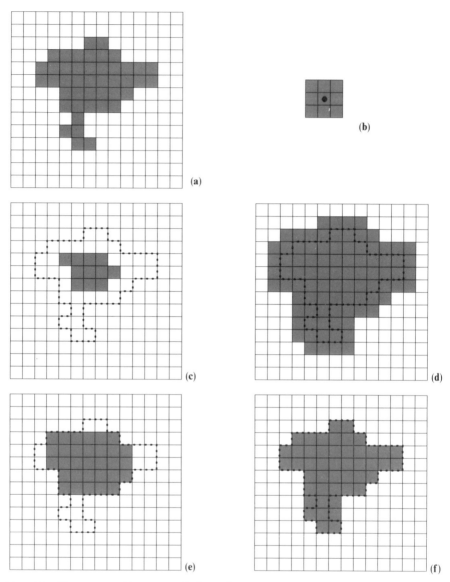

Fig. 5.5 Illustration of basic morphological operations: (a) original set, extracted as a subset from the turbinate image; (b) structuring element: a square of side 3 (the reference pixel is at the centre); (c) erosion; (d) dilation; (e) opening; (f) closing

that is, the set of positions (i, j) for which $S^1_{(i,j)}$ is contained in A and $S^2_{(i,j)}$ lies completely outside A. It follows that for the transform to make sense, $S^1_{(i,j)}$ and $S^2_{(i,j)}$ must not overlap. It can be seen that an erosion is a special case of a hit-or-miss transform, with $S^2_{(i,j)} = \emptyset$ (the empty set).

Operations of erosion, dilation, opening, closing and hit-or-miss can extract a wide range of types of information about a binary image and the components it contains, as well as transforming the image to remove or retain objects satisfying some criteria regarding their structure. What aspects of sets we look at depends on the structuring elements used. Discs or squares of different sizes interact with the size and shape of objects. These are considered in §5.2. Hit-or-miss transforms that interact with the boundary of sets give information on connectivity properties (§5.3), and structuring elements consisting of pairs of separated pixels are affected by the texture of the sets under study (§5.4).

5.2   OPERATIONS FOR SIZE AND SHAPE

The most straightforward morphological operations, and the easiest to use, are those relating to the size and local shape properties of objects in images. For this, we use structuring elements that are discs or squares. Image components, or their background, will be affected according to whether they can accommodate discs or squares of the size used. Figure 5.6 shows the effect of a closing with a lattice approximation to a disc of radius 3 pixels on the black part of the turbinate image. This could equally be regarded as an opening of the white part of the image. We see that it has had the effect of removing small black speckly features in the white area of the image, when compared with the original, and is akin to the manual cleaning up shown in Fig. 1.2(b). Although it has not accomplished all the cleaning desired, it has done much of it.

(a)   (b)

Fig. 5.6   (a) The turbinate image. (b) The turbinate image after closing with a disc of radius 3 pixels.

Discs are a natural choice as structuring element when we are interested in size criteria, since they are *isotropic* (the same in all directions). It is possible to combine operations involving discs. If we denote the disc of radius r as B_r then the following properties hold, where X denotes any image or set:

- Eroding with two discs, one after the other, is equivalent to eroding with one big disc, whose radius is the sum of the radii of the smaller discs:

$$(X \ominus B_r) \ominus B_s = X \ominus B_{r+s}.$$

- Opening with two discs is equivalent to opening with the larger disc only:

$$\psi_{B_r}(\psi_{B_s}(X)) = \psi_{B_{\max(r,s)}}(X).$$

There are analogous properties for dilation and closing respectively. Similar properties can be stated for L_r, the square of size $(2r + 1) \times (2r + 1)$.

The above properties are useful in constructing efficient algorithms for performing operations with large structuring elements. For integer k, the erosion $X \ominus B_k$ is equivalent to $X \ominus B_1 \ominus B_1 \ominus \cdots \ominus B_1$ (k times). However, the exact equivalence of these operations breaks down on the lattice, since no lattice disc corresponds exactly to a disc in continuous space (unlike the square, for which the equivalence *is* exact). Figure 5.7 shows the differences between discs of radius 2, 4, 6 and 8 and what is obtained when they are produced by multiple dilations of the disc of radius 2. (Discs of radius r are defined as the set of lattice points whose distance from the centre pixel is $\leq r + \frac{1}{2}$.)

Another useful approach for implementing operations with discs of large radii is the *distance transform*. This replaces the pixel values in one part of a binary image (say the pixels labelled 0) with their distance to the nearest pixel labelled 1. It is then straightforward to perform an erosion with a disc of radius r simply by removing all pixels whose distance label is less than r. A dilation is similarly performed by eroding the background. Forming an exact distance transform is computationally intensive, but efficient algorithms that form good approximations to distance transforms are available (Danielsson, 1980; Borgefors, 1986). One easy-to-program and reasonably accurate algorithm is as follows:

The algorithm operates in two passes through the image. We start with the binary image, and assume we want to perform the transform on the black part. We shall record for each black pixel where the nearest white pixel is. For pixel position (i, j), denote this by (y_{ij}, x_{ij}). Let its distance from (i, j) be d_{ij}, so that

$$d_{ij} = \sqrt{(i - y_{ij})^2 + (j - x_{ij})^2}.$$

This is *Euclidean* distance—see §6.1.2 for other distance measures. On the *first pass*, we visit each pixel (i, j) in turn, starting at the top left and moving along each row in turn in a raster scan (§4.2).

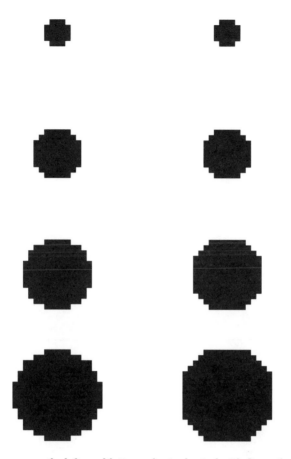

Fig. 5.7 The images on the left are dilations of a single pixel with discs of radius (from top to bottom) 2, 4, 6 and 8. The images on the right are obtained by repeatedly dilating with a disc of radius 2.

- If $f_{ij} = 1$ then $d_{ij} = 0$ and $(y_{ij}, x_{ij}) = (i, j)$.
- If $f_{ij} = 0$ then (y_{ij}, x_{ij}) becomes whichever of the locations $(y_{i-1,j-1}, x_{i-1,j-1})$, $(y_{i,j-1}, x_{i,j-1})$, $(y_{i-1,j+1}, x_{i-1,j+1})$ and $(y_{i-1,j}, x_{i-1,j})$ is nearest to (i, j). d_{ij} becomes the distance from (i, j) to (y_{ij}, x_{ij}). Note that $(i - 1, j - 1), (i, j - 1), (i - 1, j + 1)$ and $(i - 1, j)$ will already have been visited. Ties may be resolved arbitrarily.

On the *second pass*, we start at the bottom right and move towards the top left going right to left along each row in turn, performing the same steps, but re-setting (y_{ij}, x_{ij}) to whichever of (y_{ij}, x_{ij}), $(y_{i+1,j}, x_{i+1,j})$, $(y_{i+1,j-1}, x_{i+1,j-1})$, $(y_{i,j+1}, x_{i,j+1})$ and $(y_{i+1,j+1}, x_{i+1,j+1})$ is nearest to (i, j), the first of these having been obtained during the first pass. At each step, d_{ij} is re-evaluated.

If any black pixels lie on the image boundary then we shall need to define (y_{ij}, x_{ij}) for pixels beyond the boundary. For example, we may set them to be at a great distance (e.g. $(y_{0j}, x_{0j}) = (-1000, -1000)$ for all j), in which case the distance for boundary black pixels will be to the nearest white pixel in the image. Variations on the above algorithm, and a discussion of accuracy, may

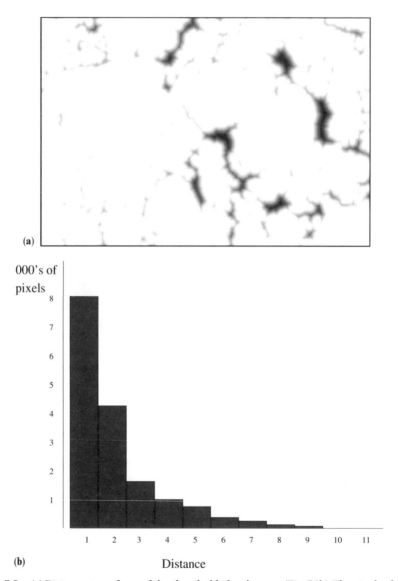

Fig. 5.8   (a) Distance transform of the thresholded soil image (Fig. 5.1b). The pixel value is proportional to the distance the pixel is from the nearest pixel shown as white in (a). Note that larger distances/pixel values are here shown *darker*. (b) A histogram of the distances.

be found in Borgefors (1986). These include algorithms based on other definitions of distance on a lattice (see §6.1.2), which are quicker to compute,
although possibly less useful in practice.

Figure 5.8(a) shows this approximate distance transform on the thresholded soil image after a closing with a disc of radius 1 (Fig. 5.1b). The transform has been applied to the dark part of the image, which represents the soil
pores. In addition to being of use in morphological erosions and dilations, the
distribution of distances can be interpreted in terms of the size distribution of
the pores. This is of relevance in studying the movement of gas and microbes
through the pore space. For example, we can see what pore space can accommodate microbes of a given size. Figure 5.8(b) shows a histogram of the distances. This distribution can also be used to develop a model of the pore space
in order to study its properties in three dimensions (Glasbey *et al.* 1991). Distributions such as this, or those based on openings of different sizes, measure
properties of the size distribution of objects in an image. This is known as *gran-
ulometry*. The morphological study of size is described by Serra (1982, Ch. X).
Related to the distance transform is the *erosion propagation* algorithm, which
allows us to associate edges with particular objects, even when they overlap.
For details and an example, see Banfield and Raftery (1992).

Alternating sequential filters

Openings and closings may be combined when we require an operation that
smooths both parts of a binary image. The simplest such combination is an
opening followed by a closing, or vice versa: $\phi_S(\psi_S(A))$ or $\psi_S(\phi_S(A))$. It can be
useful to apply a sequence of such operations, with structuring elements of
increasing size. This is because small speckle features, which would interfere
with sizing operations with the larger structuring elements, are removed by
the smaller structuring elements. If we denote the operation $\psi_k\phi_k\psi_{k-1}\phi_{k-1}\cdots$
$\psi_1\phi_1 X$ as $G_k(X)$ for integer k, where ψ_k and ϕ_k denote opening and closing with
structuring elements of size k, then

$$G_m(G_k(X)) = G_{\max(m,k)}(X),$$

with the special case $m = k$ implying idempotence. Similar properties apply to
the operation that begins with an opening. Although these combinations of
opening and closing have attractive properties, any combination that seems to
suit a purpose can be used. Figure 5.9(a) shows the X-ray image, thresholded
to select pixels in the range -69 to 100 Hounsfield units (see §1.1.3), which
are displayed as black. Figure 5.9(b) shows the effect of an opening with a disc
of radius 2 followed by a closing with a disc of radius 1. This was found to be
the most effective for the purpose intended—the measurement of the eye muscles towards the bottom of the image. This measurement is described in §6.1.

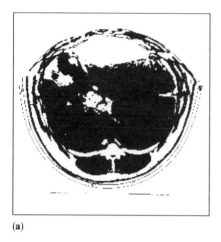
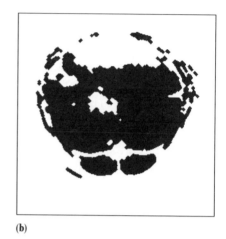

(a)   (b)

Fig. 5.9   (a) The X-ray image, thresholded to select pixel values between −69 and 100. (b) The result of opening with a disc of radius 2 followed by closing with a disc of radius 1.

We conclude this section by noting that openings and closings with elements other than discs or squares can be useful. For example, it is possible to preserve or remove elongated objects in an image by using a structuring element that is a line segment. If the objects have varying orientations, it will be necessary to combine the results of using this structuring element at several orientations.

5.3   OPERATIONS FOR CONNECTIVITY

Often it is necessary to consider the connectivity of the objects in an image. This is something that is not taken into account in the sizing operations described in §5.2, in which the operators did not interact with the total size of an object, but only its 'local' size, which is an aspect of the shape of the object. If we wish to perform operations pertaining to a connected object as a whole, a different approach is needed. Morphological operations can be used to handle much of this, although in some cases a more direct approach can be more efficient. Note that in dealing with connectivity, we can choose to assume either 4-connectivity or 8-connectivity (§4.2). Unless stated otherwise, we shall assume 8-connectivity.

Counting the number of connected components in an image is a useful operation. Although there are morphological operations to do this efficiently, the simplest approach is to label the connected components individually, as described in §4.2. An immediate by-product of this algorithm is a count of the number of connected components.

A morphological tool that does not change connectivity is that of *conditional dilation*, sometimes called *geodesic dilation*. This involves dilating a set and performing an intersection (Fig. 5.2(f)—sometimes termed an *'and'* operation) with another set. For example, if we perform an erosion that completely removes objects which do not satisfy some shape criterion, we shall also have changed the shape of the objects that have not been eliminated. They can be restored by a sequence of dilations constrained to lie *within* the original image. If I is the original image, and J_0 the result of an operation that has completely removed unwanted objects but left parts of all other sets, then the iteration using L_1, the 3×3 square with its reference pixel at the centre,

$$J_{n+1} = (J_n \oplus L_1) \cap I$$

repeated until $J_{n+1} = J_n$, will produce a result in which the unwanted objects are removed, and the remaining objects are reconstructed. However, it is often easier to work with labelled objects. In the above case, we should simply need to keep track of the labels of the eroded objects, and use this information to filter the image of labels, so that only certain objects remain.

Conditional dilation also allows us to perform other useful image operations. For example, if we wish to remove objects that touch the boundary of the image then a conditional dilation starting from the set of all black pixels at the boundary will find all pixels in boundary-touching objects. A conditional dilation of the white part of the image, starting from the white boundary pixels, will fill the image background, but not holes in black objects. This can then be used to fill such holes, if this is desired.

Thinning and thickening

The size and shape operations in §5.2 do not preserve connectedness. We may be concerned not simply with whether or not an object is a single connected component, but with properties such as how many branches and holes it contains. Such properties are referred to as the *topology* of the object. We can ask that this topology be preserved. Operations that do this are termed *homeomorphisms*, and they may be described as *homotopic*. They are known as *thinning* and *thickening* operations, depending on whether they add to or remove black pixels from the image.

The general definition of a thinning of a set A by a structuring element S is that we remove from A a part of A specified by the hit-or-miss transform $A \circledast S$. The thinning is denoted by $A \bigcirc S$ and may be written in set notation:

$$A \bigcirc S = A \setminus (A \circledast S),$$

where the set operation $X \setminus Y$ is a subtraction resulting in the elements that are in X but not in Y, i.e. $X \cap Y^c$. To use the hit-or-miss transform $A \circledast S$, the two components of S must be chosen so that connectivity is unaffected.

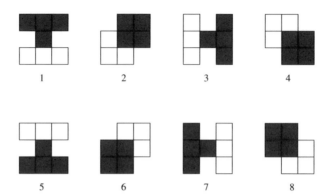

Fig. 5.10 Structuring elements for thinning. These are four orientations of each of two similar structuring elements. The shaded squares are elements of S^1 (i.e. they must be contained in the object), and the empty squares are the elements of S^2, which must lie outside the object. The central pixel is the reference pixel. The structuring elements are used in the order $1, 2, \cdots, 8$.

Thinning is most often used to reduce objects to a thickness of one pixel. Three features of the way this is done are as follows:

- A set of rotations of a basic structuring element is applied in sequence.
- The thinning is applied repeatedly until no further change occurs.
- Sometimes it is desirable to *prune* the thinned objects to remove small barbs that are not considered to be part of the true structure of the object.

Figure 5.10 shows some standard structuring elements for thinning. We may think of these structuring elements as eating away at the boundaries of a set without breaking local 4-connectedness or introducing holes inside the set. An algorithm to perform this thinning on an image would apply the hit-or-miss transform with the sequence of structuring elements in Fig. 5.10 in turn. The sequence would then be repeated until no change was found to occur. To thin the black part of an image with the first structuring element in Fig. 5.10 would mean setting

$$
g_{ij} = \begin{cases} 1 & \text{if} \quad \begin{aligned} & f_{ij} = 0, \\ & f_{i-1,j-1} = 0, \quad f_{i-1,j} = 0, \quad\quad f_{i-1,j+1} = 0, \\ & f_{i+1,j-1} = 1, \quad f_{i+1,j} = 1 \quad \text{and} \quad f_{i+1,j+1} = 1, \end{aligned} \\ f_{ij} & \text{otherwise} \end{cases}
$$

The second structuring element is then used to thin g, etc. Figure 5.11(b) shows the result of applying them in sequence to a thresholded subset of the fungal image until no further change occurs. The position and length of the hyphae have been unaffected, but they are now only one pixel thick. A

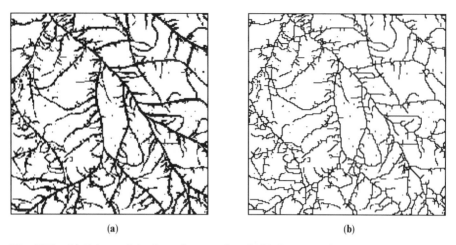

(a)   (b)

Fig. 5.11   (a) Subset of the fungal image thresholded at a pixel value of 150. (b) Same image after applying thinning operations of Fig. 5.10 until no further change occurs.

different set of operations could be used to thin to a minimal 8-connected set of pixels. The above is just one of many algorithms that have been proposed for thinning. For a discussion of some others, see Rosenfeld and Kak (1982; Ch. 11), Chen and Hsu (1990) and Jang and Chin (1990).

The thickening operation $A \odot S$ is defined by

$$A \odot S = A \cup (A \circledast S)$$

If S is symmetrical with respect to a set and its complement then thickening is the equivalent of thinning the complement of the set. One application of this is that by thinning and pruning the (white) background to a thickness of one pixel, we shall have found that part of the background around each black object that is, in some sense, more associated with that object than with other black objects. This is known as the skeleton of influence zone or *SKIZ*. For more details, see Serra (1982, Ch. XI).

The thinned image above is termed a *skeleton* of the original image—it follows the shape of the image, but is only one pixel thick. This can be a useful transform of the image. For example, with the fungal image, it enables us to estimate the total length of the fungal hyphae by counting the number of black pixels in the skeleton image. (We shall need to adjust this for the different lengths of digital distance in the horizontal, vertical and diagonal directions—see §6.1.2.) There are other possible definitions of the skeleton of an object. Our basic requirements would be that

- it preserves the topology of the object;
- it is one pixel thick;
- it is in the 'middle' of the object.

The third point could be achieved in various ways. For example, we could ask that any pixel on the skeleton is equidistant from two non-adjacent pixels on the boundary of the object. Such a skeleton is termed the *medial axis*.

The thinning defined above satisfies the first two of these properties, but is only an approximation to the medial axis. Other approaches to obtaining a skeleton are also possible. For example, Kalles and Morris (1993) divide the image into trapezoidal blocks and generate appropriate skeletons for each block, which may then be combined. Other methods are suggested by Arcelli and Sanniti di Baja (1989) and Xia (1989). A discussion of a number of other morphological approaches to defining and constructing the skeleton in a digital image may be found in Serra (1982, Ch. XI) and Meyer (1989). A discussion of the relevance of the skeleton and related concepts in studying the shape of biological objects may be found in Blum (1973).

A skeleton can be pruned by looking for end pixels (i.e. pixels in the skeleton with only one neighbouring pixel in the set), and removing those joining another branch of the skeleton within a specified number of pixels along the thinned object. A branch junction can be recognized as one where a pixel has three neighbouring pixels. Pruning can be regarded as a form of thinning, since it involves removing pixels on the basis of some criterion. Pruning is useful if the skeleton is affected by small features that are not of interest or are due to noise. A complete pruning—removing all branches of the skeleton that come to an end point—can be used if we wish to preserve only closed loops in a skeleton. (See Serra (1982, Ch. XI).)

5.4   OPERATIONS FOR TEXTURE

Texture is a concept for which it is difficult to give an exact definition. In one approach (sometimes termed syntactic pattern), it is taken to mean the spatial arrangement of features of an image, and is used when there is some pattern (another difficult word to define!) in this arrangement—when there is something not totally random. These ideas are relevant only where there are features that are repeated in the image. For example, there are several cells in the algal and muscle fibres images, many bands in the DNA image and many hyphae in the fungal image. On the other hand, the features of the X-ray, turbinate and chest images are not repeated, and so the concept of texture as defined above does not arise in these examples.

The most useful and simplest tool for studying the texture of a binary image is erosion by a structuring element consisting of two pixels a specified distance apart, followed by a count of the pixels remaining after doing this. This may be repeated for a number of different distances and orientations. The number of pixels remaining after erosion, as a function of distance, summarizes the texture of the binary image. This function may be termed the *auto-crossproduct* function of the image, since it records the number of times pixels a given

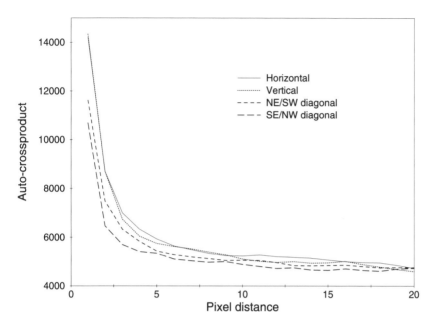

Fig. 5.12 Auto-crossproduct function for the thresholded fungal image for four directions: horizontal, vertical and the two diagonals.

distance apart have a product of 1 (i.e. are both 1). (Serra (1982, p. 272) uses the term 'covariance' for this function, although this differs from the usual statistical use of this word.) If the texture is not isotropic, the orientation of the structuring element will be important. The variation in the auto-crossproduct, as a function of the direction of the structuring element, may be used as a description of the *anisotropy* (variation with direction) of the texture.

Figure 5.12 shows the auto-crossproduct function for the thresholded fungal image for four directions: horizontal, vertical and the two diagonals. There is little evidence of any periodicity, which would be shown by a peak in the function other than at zero. There is also little evidence of any anisotropy. Figure 5.13(a) shows a thresholded section of the DNA image, and Fig. 5.13(b) shows the auto-crossproduct function for the four directions. Here we see that there is evidence of periodicity in the vertical direction, but not in the horizontal direction. This is a consequence of the regular horizontal banding structure in parts of this image.

5.5 MORPHOLOGY FOR GREYSCALE IMAGES

The morphology discussed so far was developed for binary images. Morphological techniques are most often used for such images. Morphology is best used with low-noise images, which can often be thresholded with little loss of

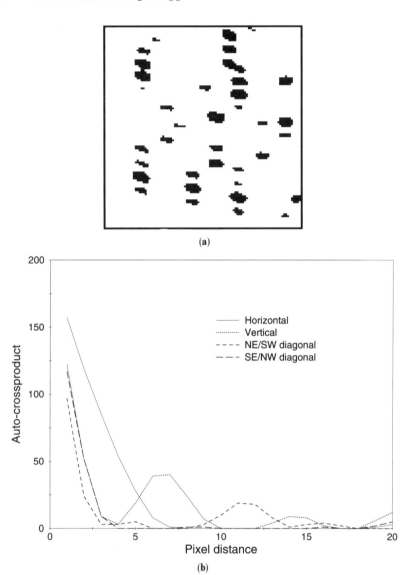

(a)

(b)

Fig. 5.13   (a) Section of DNA image, thresholded at 170. (b) Auto-crossproduct function for four directions: horizontal, vertical and the two diagonals.

information. Where the image consists of a few distinct grey levels (or, in practice, a few clusters of grey levels) then analysis of the binary images obtained with a few different thresholds may be appropriate. This could be used with the X-ray image, for example.

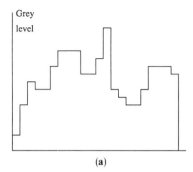 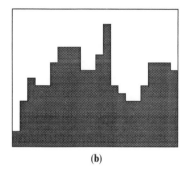

(a)   **(b)**

Fig. 5.14   (a) A one-dimensional function. (b) Equivalent two-dimensional binary image.

The most general way of handling a greyscale image is to regard it as a binary set in three dimensions. We imagine the image as defining a function in the two-dimensional plane, which produces a map in the third dimension (as in Fig. 1.5). The black part of the image is the volume under the function (sometimes termed the *umbra*), and the white part, its complement, is the volume above the function. This is analogous to using a one-dimensional function to define a set in two dimensions, as illustrated in Fig. 5.14. The morphological properties of the two-dimensional set are determined by the features of the one-dimensional function, and so these can be studied by operations on the two-dimensional set. An analogous situation applies to two-dimensional functions and three-dimensional sets, and this is how we approach morphology for greyscale images.

For mathematical morphology in three dimensions, we simply need to consider three-dimensional structuring elements. The only part of the umbra relevant for morphological operations is the top, since the rest is always the same—it continues downward indefinitely. (Similarly, for the one-dimensional function in Fig. 5.14, we need only be concerned with the top of the two-dimensional set. The rest provides no information). Similarly, only the top of the structuring element is important, since this is the part that will interact with the top of the umbra. Thus the structuring element can also be considered as a function in two dimensions. Commonly used functions are those that are constant over a region, particularly a square or disc (so that the structuring element is a cuboid or cylinder) or are like a hemisphere, so that the structuring element can be thought of as a sphere. A cone is another possibility. Some structuring elements are illustrated in Fig. 5.15.

The complement of the binary set we have defined is the region *above* the function. In Fig. 5.14(b), this corresponds to the white part of the image. All operations on the volume below the function correspond to a complementary operation on the volume above the function. The area above the function can be thought of as the binary set corresponding to the negative (in the photographic sense) of the image.

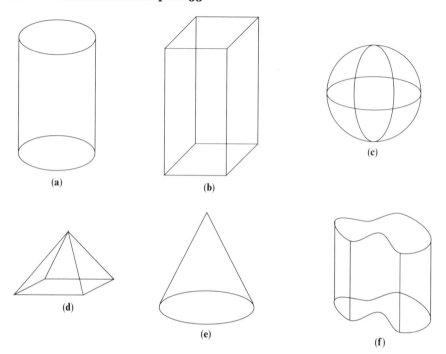

Fig. 5.15   Three-dimensional structuring elements: (a) cylinder; (b) square cuboid; (c) sphere; (d) square pyramid; (e) cone; (f) arbitrary flat-topped structuring element.

Erosion, dilation, opening and closing

As in binary morphology, the operations of erosion, dilation, opening, closing and hit-or-miss play a central role. They are straightforward to obtain, and it can be shown that if the structuring element is flat-topped (as in Fig. 5.15f), the top surface of which is a 2D set S (which we also use to denote the structuring element itself), then greyscale erosion and dilation are as follows:

$$(f \ominus S)_{ij} = \min_{(k,l)\in S} f_{i+k,j+l},$$

$$(f \oplus S)_{ij} = \max_{(k,l)\in S} f_{i+k,j+l},$$

where $(k,l) = (0,0)$ is the reference pixel in S. Thus we are transforming f by taking the minimum or maximum of f in a neighbourhood, corresponding to the region S, about each pixel of f. (We also used minimum and maximum filters in §3.4.1.) Fast algorithms for minimum and maximum filters in square or octagonal regions are discussed by van Herk (1992). Openings and closings are defined as before, so that $\psi_S(f) = f \ominus S \oplus S'$ and $\phi_S(f) = f \oplus S \ominus S'$. It can be shown that these operations are equivalent to performing operations with the corresponding two-dimensional structuring elements simultaneously on all possible thresholded versions of the image, and reconstructing a greyscale image from these. When S is a square cuboid or a cylinder, operations with the structuring element S can be decomposed as seen in §5.2. The sequences of alternating openings and closings described in §5.2 can also be used. For an example, see Meyer (1992).

Figure 5.16 shows the effect of applying these operations to the cashmere image using a cylinder of radius 6 as the structuring element. The erosion (Fig. 5.16a) replaces each pixel with the minimum in a disc of radius 6 centred on that pixel. Since the fibres are darker than their background, or

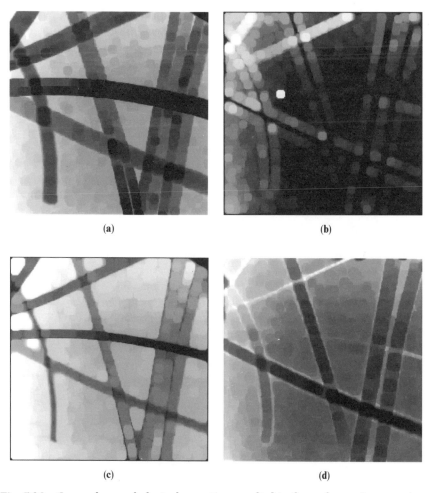

(a) (b)

(c) (d)

Fig. 5.16 Greyscale morphological operations applied to the cashmere image, using a cylinder of radius 6: (a) erosion; (b) dilation; (c) opening; (d) closing.

have edges that are darker, this has led to each fibre appearing darker and wider than in the original image. The bright borders on some fibres have disappeared, and the fibres that are brighter in the centre no longer have this feature since the pixels in the centre are within 6 pixels of a dark edge. The dilation (Fig. 5.16b) replaces each pixel with the maximum within a distance of 6 pixels. The bright edges of some fibres are now thicker, and in some cases the whole fibre now appears bright. Some fibres have almost disappeared. The opening (Fig. 5.16c) is similar to the erosion for this image, but the fibres, while still mostly dark, have now been shrunk back to their original size. The closing (Fig. 5.16d) has produced a mixture of thick dark fibres and thin bright fibres.

The effect of using a spherical structuring element is demonstrated in Fig. 5.17. Comparing its effect with that of a disc is a little like comparing a moving average filter with a Gaussian filter where the weights decay away from the centre. When opening with a sphere, the effect of nearby pixels decreases away from the centre pixel. This can be seen geometrically by thinking of opening with a cylinder as trying to fit a cylinder into the bumps of the function (from below) and similarly with a sphere. Opening or closing with a

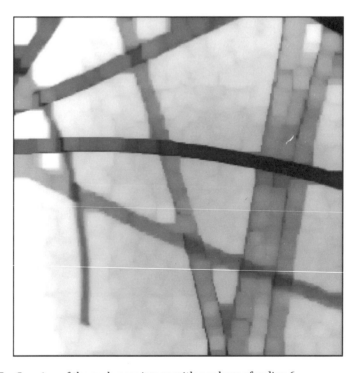

Fig. 5.17   Opening of the cashmere image with a sphere of radius 6

sphere is sometimes known as a *rolling ball* transform. The differences between Figs 5.16(c) and 5.17 are very subtle. For example, the intersections between the edges of two fibres are less rounded in Fig. 5.17. A sphere can be rolled more tightly into the corners of the bright parts of the image between the fibres than a disc can slide. Fast algorithms for erosion and dilation using spheres or discs are described by Adams (1993).

Erosion and dilation with a general structuring element require relatively simple algorithms, but are calculated rather slowly. If the top of the structuring element can be represented by a function g_{kl}, (without loss of generality $g_{00} = 0$), then it follows that

$$(f \ominus g)_{ij} = \min_{(k,l)} \ (f_{i+k, j+l} + g_{kl}),$$

$$(f \oplus g)_{ij} = \max_{(k,l)} \ (f_{i+k, j+l} + g_{kl}),$$

where (k, l) ranges over the domain of definition of g. A discussion of the use of other structuring elements and of iterated operations may be found in Sternberg (1986). An example of combining operations is described by Skolnick (1986). The use of conditional dilations to extract features satisfying given conditions, and efficient algorithms for doing this, are described by Vincent (1993).

Morphological erosion, openings etc. are often used to smooth an image, and if so they can be thought of as types of nonlinear filters. Their effect on the cashmere image is to change the thickness or remove the bright and dark lines associated with the fibres. We may use this to generate an estimate of the fibre diameter distribution, as illustrated in Fig. 5.18 (colour plate). Figure 5.18(a) shows the eroded image (Fig. 5.16a) thresholded at a pixel value of 140. Figure 5.18(b) shows the effect of a *sizing transform* on the black part of the thresholded image. This places discs of maximum radius in the black part of the image—the pixel values are the radius of the largest disc that can be placed in the black part of the image and contain that pixel position. Figure 5.18(b) is shown in pseudocolour. The sizing transform may be obtained in a number of ways. One is to perform openings with discs of increasing integer radii until none of the black part of the image remains. The largest radius at which a pixel remains after the opening is its value in the sizing transform. Figure 5.18(c) shows a histogram of the pixel values in Fig. 5.18(b). Each count was divided by the pixel value to produce a length-weighted, rather than an area-weighted, distribution. In order to use this to estimate average fibre diameters, we shall need to remove large values (where fibres coalesce) and small values (due to noise). We shall also need to shrink each size value to compensate for the expanding effect of the erosion in Fig. 5.16(a). An alternative, which did not work as well for this image, would be to use the opening in Fig. 5.16(c). If these difficulties are dealt with, we should be able to use this approach to measure fibre diameters automatically.

The top-hat transform

Openings and closings can be used to derive useful image operations that are not themselves morphological (in the sense of being expressible as hit-or-miss transforms). The range filter, the difference between a greyscale dilation and erosion, is described in §3.4.1. Another one is the *top-hat* transform.

The top-hat transform is used for extracting small or narrow, bright or dark features in an image. It is useful when variations in the background mean that this cannot be achieved by a simple threshold. If C denotes a cylinder (whence the term 'top hat') with radius r then subtracting the closing with C from the original image, i.e. taking the transform

$$f - \phi_C(f)$$

will find narrow dark features in the image. This is because the closing will have eliminated them, and they will be apparent when the closing is subtracted from the original image. For finding bright features, the transform $f - \psi_C(f)$, i.e. the subtraction of the opening, is appropriate. Figure 5.19 shows the effect of the top-hat transform (subtracting the closing) on the DNA image. This image shows how dark the bands are relative to the local background. This has removed the trend in the original image in which the background is darker near the top, and in Fig. 5.19 (which has been contrast stretched using the minimum and maximum pixel values) the bands stand out more clearly. A disc of radius 4 was used. The radius of the disc should be chosen on the basis of the size of features to be extracted.

Greyscale texture

Greyscale texture is more complex to consider than binary texture. Many approaches, both morphological and otherwise, have been proposed. A review may be found in Reed and du Buf (1987). Morphological approaches may be based on generalizations of the structuring elements discussed in §5.4. The basic element of two pixels can now have an angle in the third dimension, as well as an orientation. Often, it is simpler to study the texture of a thresholded version of a greyscale image, as was done with the DNA image. If this is an oversimplification, several thresholds may be used to build up a more complete picture.

Greyscale topology

When an image is represented as a function in the two-dimensional plane (as in Fig. 1.5), with the pixel values representing height above the plane, it can be

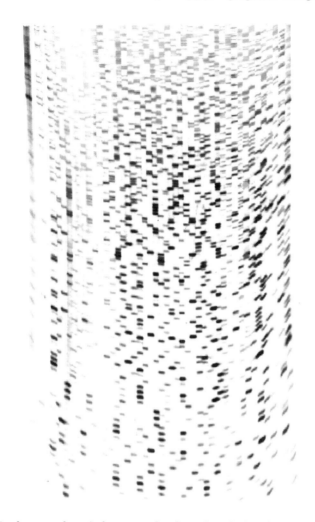

Fig. 5.19 Top-hat transform (subtracting the closing) applied to the DNA image. A disc of radius 4 was used.

useful to look for features of the resulting 'landscape'. For example, bright regions in the image can be thought of as hills with the local maxima of brightness being summits or peaks. Similarly, dark regions in the image give rise to channels, depressions and pits. We can go further and define concepts such as ridges, valleys, watersheds and divides in the 'topography' of an image. These sometimes correspond to features of interest in the image under study, and recognizing and extracting them can be a useful technique.

Some ideas based on greyscale topology, such as local minima and watersheds, have already been discussed in §4.3. We shall not pursue them further here, except to note that morphology often provides an effective method of describing topological features, and finding them in images. Further details may be found in Serra (1982, Ch. XII) and Meyer (1992). Some efficient (although complex) algorithms are discussed in Bleau, Deguise and Leblanc (1992). Useful discussions may also be found in Haralick, Watson and Laffey (1983) and Haralick (1983).

5.6 SUMMARY

The main points of this chapter are as follows:

- Mathematical morphology, which is based on set theory, provides useful tools for image operations.
- Morphology is complementary to the other methods described in this book.
- Morphology has most often been used with binary images, but is now also used with greyscale images.
- A fundamental operation is that of erosion, which looks at where a test set can fit into the image.
- From the erosion we can also define
  — dilation, which is an erosion on the complement of a set;
  — opening, an erosion followed by a dilation;
  — closing, a dilation followed by an erosion.
- The hit-or-miss transform is the most general morphological operation, and is based on a structuring element with two components. The transform involves looking for pixel positions where one component lies within the set of black pixels, and the other lies completely without.
- Operations based on square or circular structuring elements are the most commonly used. They interact with the size of features in the image.
- A distance transform can provide an efficient way to perform morphological operations, and can also be of interest in its own right.
- Morphological thinnings and skeletons preserve the topology of objects while reducing them to single pixel thickness.
- Texture may be studied using structuring elements consisting of pixels separated by a distance.
- Morphology may be extended to greyscale images by considering them as binary images in three dimensions.
- A variety of three-dimensional structuring elements may be used, and erosion, dilations, openings and closings may be defined. They are used to smooth an image, or to help extract particular structures.
- The top-hat transform enables us to extract small or thin, bright or dark objects from a varying background.

6

Measurement

The extraction of quantitative information from images is the endpoint of much image analysis. The objective may simply be to

- *count the number* of objects in a scene,
- compute their *areas*, or
- measure *distances* between objects.

Alternatively, it may be necessary to

- characterize the *shapes* of objects, or
- summarize their *boundaries*,

in order to discriminate between the objects or to summarize information about them.

To illustrate the range of techniques to be covered, let us consider Fig. 6.1(a). This shows regions of muscle extracted from the X-ray CT image of a sheep (Fig 1.7(c)). These regions of eye-muscle, or *longissimus dorsi*, the muscles that lie along a sheep's back, were obtained by thresholding the X-ray image and then using morphological openings and closings to reduce the roughness in the boundaries (see Fig. 5.9b).

What measurements could we take from these regions?

- We could compute the cross-sectional areas of the muscles. These measures of size can be obtained simply by counting the number of pixels in both black regions in Fig. 6.1(a). The counts are 1013 and 970 pixels for the left and right muscle respectively. Because 1 pixel is equivalent to an area of 1 mm^2 in this application, these counts convert to areas of 10.1 and 9.7 cm^2 for the two muscles.
- Alternatively, we may wish to quantify the shapes of the muscle cross-sections, by using a measure of *elongation*, for example. One definition of this measure is the ratio of the *length* (that is, the maximum distance between two pixels within a muscle) to the *breadth*. The line segments that correspond to the lengths are shown in Fig. 6.1(b). There are several ways to define the breadth: the one we have used is the perpendicular distance from one of these lines to the furthest pixel on one side of the muscle, *plus*

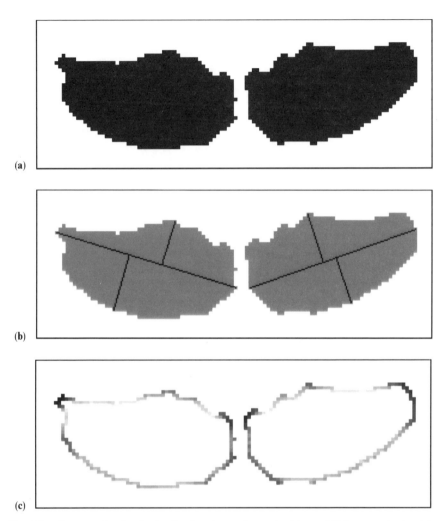

Fig. 6.1 Eye muscles in the back of a sheep, from the X-ray CT image, obtained by thresholding the original image followed by morphological opening and closing, and used to illustrate different types of image measurement: (a) to estimate cross-sectional areas; (b) to measure elongation of the muscle cross-sections, as the length-to-breadth ratios; (c) to describe muscle boundaries in terms of local curvature—dark pixels indicate a small radius of curvature and lighter pixels indicate a larger radius.

the distance to the furthest pixel on the other side. The line segments that correspond to the breadths are also shown in Fig. 6.1(b). Bear in mind that we consider pixels to be points on a lattice, although we display them as square blocks, and therefore it is appropriate to measure distances between the *centres* of the blocks. For the left and right muscles, the length-to-breadth ratios are $52.2/27.9 = 1.87$ and $48.7/26.6 = 1.83$ respectively. (Further details of how these distances were obtained are given in §6.1.2.)

- A third possibility is to describe the *boundaries* of the muscles—for example by measuring their *curvature*. Figure 6.1(c) shows the local curvature around the pixellated boundary of both eye-muscles. Darker pixels indicate a smaller radius of curvature, and lighter pixels indicate a larger radius (more details are given in §6.3).

Measurements are usually taken from the images output from segmentation algorithms (Chapter 4), which have possibly also been processed using morphological operators (Chapter 5). Such will be the case for most of the examples considered in this chapter. However, in some applications, measurements can be obtained directly from an image without any preprocessing being required, while in other cases it may be necessary to combine segmentation and greyscale information. Examples will be given in §6.1.1.

The three forms of image measurement in Fig. 6.1 illustrate the coverage in the three sections that follow. In §6.1, ways will be considered in which size measurements can be obtained from an image. Then, in §6.2, aspects of shape that are independent of size will be presented. Descriptions of the boundaries of objects will be considered in §6.3. Finally, in §6.4, the main points of the chapter will be summarized.

6.1   MEASURES OF SIZE

It is a straightforward matter to count the number of objects in an image using the connected-components algorithm in §4.2, provided that the segmentation has successfully associated one, and only one, component with each object. If this is not the case then manual intervention may be necessary to complete the segmentation, as was illustrated for the muscle fibres in Fig. 4.10(b). However, short-cuts can sometimes be taken. For example, if the mean size of objects is known, then the number of objects in an image can be estimated even when they are touching, through dividing the total area covered by all the objects by this average size. It is even possible to make allowance for objects overlapping each other provided that this process can be modelled, for instance by assuming that objects are positioned at random over the image and making use of the properties of these so-called *Boolean models* (Cressie, 1991, pp. 753–759). For example, Jeulin (1993) estimated the size distribution of a powder.

It is also important to take account of any objects that *overlap the border* of an image. If we wish to estimate average areas or perimeters of objects then the simplest solution is to ignore such objects, although this may introduce some downward bias because larger objects are more likely to intersect the border. However, if the aim is to estimate the number of objects per unit area, this will be overestimated if we include all objects in the count, but underestimated if we exclude those that overlap the border. An approximately unbiased estimate is obtained if objects only in contact with the bottom or right image borders are ignored.

It is always sound practice, if possible, to *calibrate* the results produced by image analysis. This involves obtaining a few of the required measurements by some other means that are known to be accurate, and comparing the answers with those from image analysis. This is well worthwhile, even it is time-consuming or expensive, because it secures much greater credibility for the results produced by image analysis. If there is a discrepancy between the two types of measurement then in some cases an adjustment for bias can be applied to the image analysis results.

Fowler *et al.* (1990) provide an excellent case study of calibration in MRI (magnetic resonance imaging). Accuracy of volume estimates was assessed by imaging bottles of copper sulphate at a range of positions in the imaging plane. Also, the test object that can be seen in Figs 1.7(a, b), consisting of 80% water, was used as a standard against which to estimate water content of different tissues.

The two most common types of statistics used to describe an object's size are measurements of area and of distance. Measurements of area and related moments will be covered in §6.1.1. Many variants of distance, including perimeters and diameters, will be dealt with in §6.1.2.

6.1.1   Areas and moments

It is not always necessary to segment an image in order to estimate the area of a region. For example, in the Landsat image, we may want to measure the area covered by oil-seed rape fields—these are the approximately rectangular regions that appear as yellow and yellowish-green in Fig. 2.10. Figure 6.2 shows a plot of pixel values in band 2 (green) against those in band 1 (blue). Yellow pixels have been identified manually in this scatterplot as those lying in the outlined region with a higher green than blue signal. They total 6301 pixels, which converts to an area of 5.7 km^2, because 1 pixel is equivalent to 900 m^2 on the ground. Of course, not all pixels will have been correctly classified by this criterion, but, provided the number of oil-seed rape pixels that have been missed approximately cancels out the number of other pixels incorrectly classified as rape, the overall estimate will be reasonably accurate. This will depend on both misclassification rates and total areas involved, and may be a somewhat optimistic assumption. However, the estimate can be improved and the uncertainty quantified, if the true ground cover can be determined for part of the image (for example, by visiting the region and identifying the crop growing in each field). Then a *ratio estimate* can be constructed, making use of the satellite data as an auxiliary variate (see e.g. Cochran, 1977, Ch. 6).

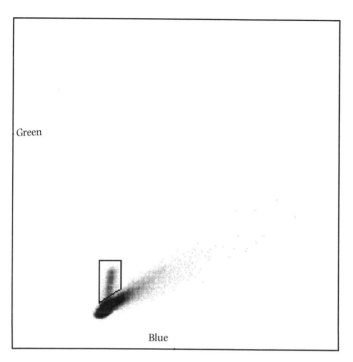

Fig. 6.2 A scatter plot of Landsat band 2 (green) against band 1 (blue), with pixel frequency displayed on a log-scale, and the range of values enclosed by the quadrilateral corresponding to pixels in oil-seed rape fields.

Let us now consider a segmented image, and let the object that we are interested in measuring be denoted by all pixel locations (i, j) such that $(i, j) \in A$. We shall use the set notation of Chapter 5 (see Fig. 5.2a). The *area* is the number of pixels in A, and can be expressed mathematically as

$$\text{area} = \sum \sum_{(i,j) \in A} 1;$$

that is, count 'one' for every pixel in A.

In some applications, we may be interested in the sum of pixel values within the region specified by A. This is given by

$$\sum \sum_{(i,j) \in A} f_{ij},$$

where f_{ij} denotes the greyscale value of pixel (i, j). For example, the cumulative sum of the pixel values within a spot on an electrophoresis gel (such as in Figs 1.9(a, b), after subtraction of the average background pixel value, can be used to estimate the volume of that protein in the sample. However, to achieve precise results, image digitization should be performed using a scientific laser scanner rather than the cheaper desktop scanner we used to digitize these gels. In another application, Fowler *et al.* (1990) used average MRI signals within selected regions, in images such as Figs 1.7(a, b), to estimate water content of different tissues.

We shall now consider how the definition of area can be generalized to encompass a class of measurements, termed 'moments', that can be used to specify the location and spatial distribution of an object. Instead of counting 'one' for each pixel in the expression above, we count '*i* raised to the power *k*, multiplied by *j* to the power *l*' if the pixel is in row *i* and column *j*. This is the definition of the (k, l)th-order moment:

$$\mu_{kl} = \sum_{(i,j) \in A} \sum i^k j^l \quad \text{for } k, l = 0, 1, 2, \ldots .$$

If *k* and *l* are both zero, we obtain the zeroth-order moment μ_{00}, which is the area, because $i^0 = j^0 = 1$.

First-order moments (i.e. $k + l = 1$) can be used to specify the location of an object. The *centre of gravity*, or *centroid* of *A* is a measure of the object's location in the image. It has two components, denoting the row and column positions of the point of balance of the object if it were represented by a piece of cardboard of the same dimensions. The two numbers can be calculated as the mean row number, and the mean column number, of pixels in *A*. They will not usually be integers. They are expressible in terms of zeroth- and first-order moments by

$$\text{centroid} = \left(\frac{\mu_{10}}{\mu_{00}}, \frac{\mu_{01}}{\mu_{00}} \right).$$

In applications where the positions of objects are important, the distribution of centroids can be studied using techniques of spatial point pattern analysis (Diggle, 1983).

Moments of higher order (i.e. $k + l > 1$) are also mainly determined by an object's location. The statistics are more useful if this information (which we already know about from μ_{10} and μ_{01}) is removed. *Central moments* are defined as above, except that the mean row position is subtracted from *i* and the mean column position is subtracted from *j*:

$$\mu'_{kl} = \sum_{(i,j) \in A} \sum \left(i - \frac{\mu_{10}}{\mu_{00}} \right)^k \left(j - \frac{\mu_{01}}{\mu_{00}} \right)^l \quad \text{for } k + l > 1 .$$

Central moments are *location invariant*—that is, two objects that are identical except for having different centroids will have identical values of μ'_{kl}, for all values of k and l. At least this is the case if the centroid changes by an integer number of pixels — for non-integer changes in location the pixel representation of an object may change, with consequent effects on the central moments. The moments can be expressed in terms of the earlier, non-central moments μ_{kl}. In particular, the second-order central moments are

$$\mu'_{20} = \mu_{20} - \frac{\mu_{10}^2}{\mu_{00}},$$

$$\mu'_{02} = \mu_{02} - \frac{\mu_{01}^2}{\mu_{00}}$$

$$\mu'_{11} = \mu_{11} - \frac{\mu_{10}\mu_{01}}{\mu_{00}}.$$

The second-order moments measure how dispersed the pixels in an object are from their centroid: μ'_{20} measures the object's spread over rows, μ'_{02} measures its spread over columns, and μ'_{11} is a cross-product term representing spread in the direction in which both row and column indices increase. They are proportional to the variances and covariance of a bivariate random variable with a distribution that is uniform over the set A.

Central moments are not rotationally invariant—they will change if an object is rotated. If orientation is an important feature of an object, as it will be in some applications, then it is probably desirable for the moments to be sensitive to it. But in other cases orientation is irrelevant, and moment statistics are more useful if they are invariant to rotation as well as to location. Again we must invoke the proviso that rotation will affect the pixel representation of an object, and therefore the invariance will hold only approximately. Hu (1962) derived *rotationally invariant moments*, the first two of which are the following functions of the second-order central moments:

$$\mu'_{20} + \mu'_{02},$$

$$(\mu'_{20} - \mu'_{02})^2 + 4\mu'^2_{11}.$$

The first of these statistics is the *moment of inertia*, a measure of how dispersed, in any direction, the pixels in an object are from their centroid, whereas the second statistic measures whether this dispersion is isotropic or directional. Note that Reiss (1991) showed that some of Hu's other results are in error.

There are alternative ways of combining the second-order moments in such a way that the resulting statistics are rotationally invariant. One such method is based on first specifying the direction in which the object has the maximum value for its second-order moment. This direction is

$$\phi = \frac{1}{2} \tan^{-1}\left(\frac{2\mu'_{11}}{\mu'_{02} - \mu'_{20}}\right) \quad \text{if} \quad \mu'_{02} > \mu'_{20},$$

and is otherwise this expression plus $\frac{1}{2}\pi$, where ϕ is measured clockwise, with the horizontal direction taken as zero, and $\tan^{-1}$ produces output over the range $-\frac{1}{2}\pi$ to $\frac{1}{2}\pi$. The direction ϕ, which is the *major axis* or *direction of orientation* of the object, has second-order moment

$$\lambda_1 = \mu'_{20} \sin^2\phi + \mu'_{02} \cos^2\phi + 2\mu'_{11}\sin\phi\cos\phi.$$

The direction perpendicular to ϕ, which is the *minor axis*, has the smallest second-order moment, of

$$\lambda_2 = \mu'_{20} \cos^2\phi + \mu'_{02}\sin^2\phi - 2\mu'_{11}\sin\phi\cos\phi.$$

These second-order moments, λ_1 and λ_2, are rotationally invariant, by definition. For a derivation, see Rosenfeld and Kak (1982, Vol. 2, pp. 288–290).

Table 6.1 shows several of these moment statistics for the segmented algal cells displayed in Fig. 1.12(f). For example, the region representing the cell labelled 1 consists of 358 pixels, is centred at location (18, 254), is oriented so that the major axis lies at an angle of 34° to the horizontal in a clockwise direction, and has major and minor second-order moments of 13 000 and 8100. Note that values of ϕ for all 13 cells lie in the range $-11°$ to $65°$. This is

Table 6.1 Moment statistics for algal cells in Fig. 1.12(f).

Cell	μ_{00}	$\dfrac{\mu_{10}}{\mu_{00}}$	$\dfrac{\mu_{01}}{\mu_{00}}$	μ'_{20}	μ'_{02}	μ'_{11}	ϕ	λ_1	λ_2
1	358	18	254	9600	11500	2300	34	13000	8100
2	547	50	262	22100	26500	4200	31	29000	19600
3	294	98	196	7700	8100	3600	43	11500	4200
4	774	109	174	50600	45900	6500	55	55200	41300
5	566	191	431	24700	27000	4600	38	30600	21100
6	306	200	254	7400	9700	3900	37	12600	4500
7	243	217	266	3900	5800	-400	-11	5900	3900
8	312	243	348	7400	8000	0	0	8000	7400
9	277	261	78	6900	5600	800	65	7300	5200
10	418	285	452	12600	17200	4100	30	19600	10200
11	387	296	100	9500	16000	2800	21	17100	8400
12	261	358	338	5600	5400	800	49	6200	4700
13	440	410	126	13300	19200	4000	27	21200	11300

because segmented cells from a differential interference contrast microscopic image tends to produce elliptical shapes with a common orientation.

If moments of high enough order are recorded from an object then they uniquely specify the object, which can be recovered precisely (Hu, 1962). However, this is of only theoretical interest, because, in practice, such an approach would be numerically highly unstable. Teh and Chin (1988) review other families of moment statistics, such as Zernike and Legendre moments. It is also possible to define moments using boundary pixels alone, by an appropriate change to the definition of set A, and using greyscale pixel values (Rosenfeld and Kak, 1982, Vol. 2, p. 287); for example, the zeroth-order grey-scale moment was introduced earlier in the section. Higher-order moments and their invariant forms can also be derived, though this is less commonly done. Teh and Chin (1986) and Mardia and Hainsworth (1989) concern themselves with the difference between moments evaluated on an integer lat-tice and those that would be produced if each pixel were regarded as a square of unit size. In particular, for second-order moments they suggest leaving μ'_{11} unchanged, but replacing μ'_{20} and μ'_{02} by

$$\mu'_{20} + \tfrac{1}{12}\mu'_{00} \qquad \text{and} \qquad \mu'_{02} + \tfrac{1}{12}\mu'_{00}.$$

However, these corrections do not make much difference, except for objects only a few pixels in size, and so we have not used them in our examples.

6.1.2 Distances, perimeters and diameters

The simplest of all distance measurements is that between two specified pixels (i, j) and (k, l). There are several ways in which distances can be defined on a lattice. The following are the three most common:

1. *Euclidean distance,*

$$\sqrt{(i-k)^2 + (j-l)^2}.$$

2. *Chessboard distance,*

$$\max(|i-k|, |j-l|),$$

the number of moves required by a chess king to travel from (i, j) to (k, l). It is also the length of the shortest chain of 8-connected pixels joining (i, j) and (k, l).

3. *City-block distance,*

$$|i-k| + |j-l|,$$

so-called because it is the distance we should have to travel between two locations in a city such as New York where the streets are laid out as a rectangular grid. It is also the number of pixels in the shortest chain of 4-connected pixels joining (i, j) and (k, l).

The chessboard and city-block distances can be computed more quickly than the Euclidean distance, because they require only integer arithmetic. However, in the main we shall use the Euclidean measure, because it accords with a geometric understanding of the objects to be quantified and is unaffected by the lattice orientation.

To illustrate a simple case of measuring distance, consider Fig. 6.3. This shows the ultrasound image (Fig 1.7d) after manual interpretation. In particular, a vertical line has been drawn between the top and bottom of the eye muscle at its maximum depth, using a computer mouse. This line has a length of 201 pixel units, which converts to an eye-muscle depth of 32 mm, because the vertical inter-pixel distance is 0.157 mm.

Another simple way in which distances are used in quantifying objects is that average distances, such as *average breadths*, may be produced. These can be obtained by dividing the area of an object by its length. To illustrate this, the subcutaneous fat layer of the sheep has been outlined manually in Fig. 6.3. The left end of the region has been located above a feature in the backbone, and the region is of fixed length, 307 pixels. We are interested in the average

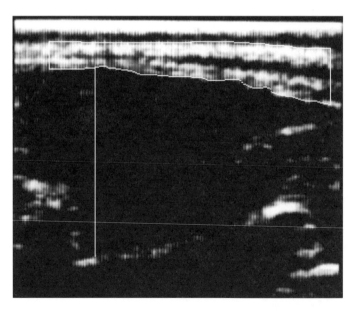

Fig. 6.3 Ultrasound image, with hand-drawn lines encircling the fat region and showing the maximum depth of the eye muscle.

fat depth. This can be measured by dividing the area of the region by the number of columns it contains, in this case $11\,987/307$, which is 39 pixel units, or 6 mm.

Curved lines

Lengths of smooth *curved lines* are sometimes required, such as the total length of fungal hyphae in Fig. 1.9(d). Before proceeding with the theory of measuring lengths of curved lines, let us think about the fungal hyphae in more detail.

We could measure the total hyphal length by dividing the area they cover by their average breadth, as we did above for the sheep fat depth. However, we do not know how wide the hyphae are, nor are they easy to measure accurately because they are only about two pixels wide. We can overcome this difficulty in a neat way by skeletonizing (§5.3) the thresholded image. Then we know that the breadth will be exactly one pixel unit. Figure 6.4(a) shows the skeleton of the thresholded image. It consists of 1121 distinct objects, with a total area of 25 838 pixels—results that were obtained using the connected-components algorithm in §4.2. If all objects less than 5 pixels in area are assumed to be the products of noise rather than hyphae and so are discarded, Fig. 6.4(b) is produced. The count is reduced to 231 components and 24 013 pixels. We have one further difficulty to contend with, namely that in order to convert the count into an estimate of length, it is necessary to consider the relationship between a line in continuous space and on a lattice.

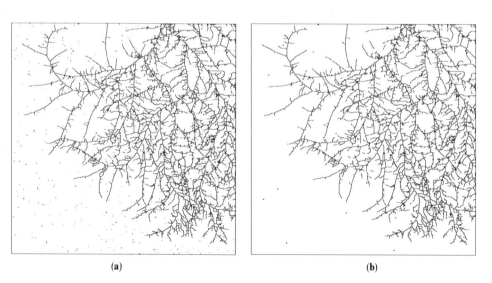

(a)   (b)

Fig. 6.4 Fungal image: (a) thresholded and skeletonized; (b) with small connected components removed.

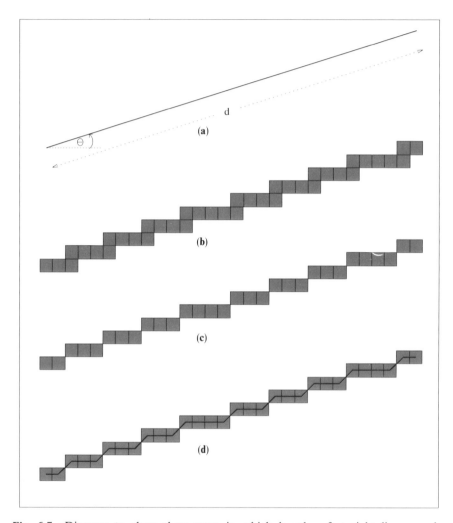

Fig. 6.5 Diagram to show three ways in which lengths of straight lines can be measured in a pixel image: (a) in continuous space the line has length d and makes an angle θ with the horizontal; (b) the 4-connected pixels lying on the line, totalling N_4, are shown; (c) the 8-connected pixels lying on the line, totalling N_8, are shown; (d) a third way of measuring the line length is by measuring the distance between diagonally adjacent pixels as $\sqrt{2}$, rather than as unity in (c).

We shall consider line lengths in general, before returning to the fungal hyphae application. Figures 6.5(b) and (c) show two possible lattice representations of the straight-line section in Fig. 6.5(a), depending on whether the pixels are 4-connected or 8-connected. We made use of connectedness in §4.2: in a 4-connected region all pixels have horizontally or vertically adjacent neighbours, whereas in an 8-connected region the neighbours may be diagonally adjacent. The line is 30.4 pixel units in length, although there are 37 pixels between the two end pixels in Fig. 6.5(b) and 28 pixels in Fig. 6.5(c). Therefore, in order to obtain good estimates of lengths in continuous space, we need to make some adjustment to pixel counts.

Consider any straight line of length d and angle θ. We need only consider θ in the range 0 to $\frac{1}{4}\pi$, because for a line at any other orientation, we could, if necessary, swap or reverse row and column labels until θ did lie in this range. The line covers $d\cos\theta$ columns and $d\sin\theta$ rows. In the 4-connected line shown in Fig. 6.5(b), $d\sin\theta$ columns have two pixels in them and the rest each have a single pixel, whereas in Fig. 6.5(c) every column has one pixel in it. Therefore the numbers of pixels are

$$d(\cos\theta + \sin\theta) \quad\text{and}\quad d\cos\theta$$

for the 4- and 8-connected cases repectively. If straight line segments at all orientations θ *occur equally often* in the curved line to be measured then, on average, the number of 4-connected pixels per unit length can be obtained by integrating over all values of θ between 0 and $\frac{1}{4}\pi$, i.e.

$$\frac{4}{\pi}\int_0^{\pi/4}(\cos\theta + \sin\theta)\,d\theta = \frac{4}{\pi} = 1.273.$$

Similarly, the number of 8-connected pixels per unit length is given by

$$\frac{4}{\pi}\int_0^{\pi/4}\cos\theta\,d\theta = \frac{4}{\pi\sqrt{2}} = 0.900.$$

Therefore a smooth curved line, which can be approximated by straight-line segments of length 1000 for instance, will be represented on the image lattice by about 1273 pixels in a 4-connected line, or 900 pixels in an 8-connected line.

Returning to the fungal image, we used a 4-connected skeleton. Also, line segment orientations were shown to be reasonably uniform in §5.4, i.e. there is no strong directional pattern in Fig. 6.4(b). Therefore, using the above result, the total length of fungal hyphae can be estimated to be 24 013/1.273 = 18 860 pixel units.

Perimeters

Perimeters of objects, that is boundary lengths, can also be measured by making use of the above theory. Let N_8 denote the number of pixels on the boundary of object A, which we count in the following way. Pixel (i, j) is on the boundary if $(i, j) \in A$, but one of its four horizontal or vertical neighbours is outside the object, that is,

$$(i + 1, j) \notin A \text{ or } (i - 1, j) \notin A \text{ or } (i, j + 1) \notin A \text{ or } (i, j - 1) \notin A.$$

Similarly, let N_4 denote the number of pixels $(i, j) \in A$ such that either one of the above four neighbouring pixels is not in A or one of the four diagonal neighbours is outside the object; that is,

$$(i + 1, j + 1) \notin A \text{ or } (i - 1, j + 1) \notin A \text{ or } (i + 1, j - 1) \notin A \text{ or } (i - 1, j - 1) \notin A.$$

The case N_4 gives the number of pixels in the 4-connected boundary of the object, whereas N_8 corresponds to the number of pixels in an 8-connected boundary. Therefore

$$\frac{N_4}{1.273} \quad \text{and} \quad \frac{N_8}{0.900}$$

are two unbiased estimators of the perimeter, provided that the assumption is satisfied that all orientations in the boundary occur equally often.

Note the following:

- The above definition of the boundary of A will also include any internal boundaries if the object has *holes*. The *Euler number* of a set is defined to be the number of connected components it consists of, minus the number of holes it contains.
- The use of scaling factors is part of *stereology*, a field that has traditionally been concerned with inference about objects using information from lower-dimensional samples—such as estimating volumes of objects from the areas of intersection with randomly positioned cutting planes (see e.g., Stoyan, Kendall and Mecke, 1987, Ch. 11). In particular, the scaling factor of 1.273 $(4/\pi)$ arises in two of the so-called 'six fundamental formulae' of classical stereology. However, the last ten years have seen a revolution in stereology, with the discovery of the *disector* (sic) and other three-dimensional sampling strategies (Stoyan, 1990).

Figure 6.5(d) shows a third way of estimating the length of a line: by measuring the distance between diagonally adjacent pixels using the Euclidean measure of $\sqrt{2}$. The number of diagonal links is $N_4 - N_8$, and the remaining $N_8 - (N_4 - N_8)$ links in the 8-connected boundary are of one pixel unit in length. Therefore the total length is

$$N_8 - (N_4 - N_8) + \sqrt{2}(N_4 - N_8) = (\sqrt{2} - 1)N_4 + (2 - \sqrt{2})N_8.$$

Repeating the previous calculations for this new expression, the expected

Table 6.2   Boundary statistics and perimeter estimates for eye-muscles in Fig. 6.1.

		Left	Right
Boundary statistics			
	N_4	160	158
	N_8	118	112
Perimeter estimators			
	$N_4/1.273$	125.7	124.1
	$N_8/0.900$	131.1	124.4
	$(0.414N_4 + 0.586N_8)/1.055$	128.3	124.2
	convex hull perimeter	128.6	123.3

number of pixels per unit length is

$$\frac{4}{\pi}\int_0^{\pi/4}\{\cos\theta + (\sqrt{2} - 1)\sin\theta\}\, d\theta = \frac{8}{\pi}(\sqrt{2} - 1) \ = \ 1.055,$$

which again can be used as a correction factor to estimate total length.

All three perimeter estimators are unbiased, provided that the condition is satisfied that all line orientations occur equally often. However, they are not all equally precise. The final estimator has the lowest variance, and should be used in preference to the other two. These, and more complicated methods for estimating perimeters yet more precisely, are considered by Dorst and Smeulders (1987) and Koplowitz and Bruckstein (1989).

Table 6.2 shows N_4 and N_8 for the sheep eye muscles shown in Fig. 6.1, together with the results obtained using the three perimeter estimators. (The convex hull perimeter will be dealt with later in the section.) Agreement is very close for the right muscle. For the left muscle, differences of about 4% have arisen because the boundary is rougher than the right one, and so the assumption that the line is smooth is not wholly appropriate.

The statistics we have considered are all measures of the *internal perimeters* of objects. If, instead, the length of the boundary of the background pixels surrounding the object were measured, this would yield the *external perimeter*, which would be greater than the internal one. The difference arises because in one case we are measuring the distances between the centres of pixels on the object side of the boundary, whereas in the other case we are measuring the distances between the centres of pixels on the other side of the boundary. A perimeter intermediate between these two can be obtained by regarding each pixel as a square of unit size, rather than as a point, and measuring the perimeter so that it goes round the boundary sides of each pixel. A short-cut to approximately the same result involves adding π to the internal perimeter. This is the extent to which the perimeter increases if an object is expanded by half the inter-pixel distance in all directions. The effect will be

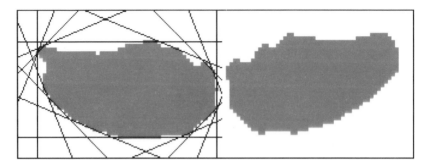

Fig. 6.6 Eye muscles from X-ray CT image, together with eight pairs of parallel tangents around the left eye-muscle at orientations that are multiples of 22.5°. They are used to estimate Feret diameters and approximate the convex hull.

noticeable only for objects a few pixels in size, and so can usually be safely ignored.

Feret diameters

The final distances we shall consider in this section are *Feret diameters*, which are defined to be the distances between parallel tangents touching opposite sides of an object. For example, Fig. 6.6 shows eight pairs of tangents to the left eye muscle at orientations that are multiples of 22.5°. At orientation θ, the Feret diameter is

$$\max_{(i,j)\in A}(i\sin\theta + j\cos\theta) - \min_{(i,j)\in A}(i\sin\theta + j\cos\theta).$$

although it is more efficiently computed using boundary pixels alone. The diameters are also referred to as *caliper diameters* because of the analogy with measuring an object's size using calipers, and as *lengths of projections* because they measure the extent of A when it is projected onto a line at orientation $\theta + \frac{1}{2}\pi$.

Certain Feret diameters are of special interest, although the terminology varies between authors.

- The *width* of an object is the range of columns it covers, i.e.

$$\max_{(i,j)\in A}(j) - \min_{(i,j)\in A}(j).$$

It is the Feret diameter at angle $\theta = 0°$.

- Similarly, the *height* of an object is the range of rows it covers, i.e.

$$\max_{(i,j)\in A}(i) - \min_{(i,j)\in A}(i).$$

It is the Feret diameter at angle $\theta = 90°$. Sometimes, the product of width and height is computed—this is the area of the enclosing rectangle.

- The maximum Feret diameter is one definition of the *length* of an object, which may alternatively be specified as

$$\max_{(i,j),(k,l)\in A} \sqrt{(i-k)^2 + (j-l)^2}.$$

The corresponding value of θ is an alternative definition of orientation to that considered in §6.1.1.

- The *breadth* of an object does not have a unique definition. Some people define it to be the minimum Feret diameter, whereas others take it to be the Feret diameter perpendicular to the length. To illustrate, Fig. 6.1(b), which has already been discussed, shows the lengths for the two eye-muscles, and the perpendicular breadths.

Length can be estimated reasonably precisely by taking the maximum of a few Feret diameters. It will be slightly underestimated, but not by much. If breadth is defined to be the minimum diameter, and is estimated in a similar fashion as the minimum of a set of diameters, then the overestimation is more severe. Therefore more Feret diameters are needed for precise measurement of the breadth. Houle and Toussaint (1988) give an efficient algorithm.

The *convex hull* of an object is defined to be the smallest convex shape that contains the object—where a convex shape is one in which if any two points within it are joined by a straight line then all the points along the line are also within the shape. The convex hull can be obtained as the region enclosed by all the tangent planes to the object, as can be seen for the left eye-muscle in Fig. 6.6. If the scale at which measurements are made is varied, by changing the magnification of a microscope for example, then the perimeter of the convex hull of the object will be found to vary less than the perimeter of the original object. In an extreme case, in which an object has a *fractal* boundary (Mandelbrot, 1982), its estimated perimeter grows infinitely large as the resolution increases. Therefore the perimeter of the convex hull can provide a more stable measure of an object's boundary length. The perimeter of the convex region defined by N equally spaced Feret diameters provides an approximation to the perimeter of the convex hull. This perimeter is expressible as

$$2N\bar{D}\tan\left(\frac{\pi}{2N}\right),$$

where $\bar{D}$ is the average of N Feret diameters at orientations that are integer multiples of π/N. Alternatively, the convex perimeter of an object can be estimated using stereological sampling methods. The convex perimeters of the eye muscles are given in Table 6.2.

Volumes of objects can be estimated, provided that something is known about their three-dimensional shape, using measurements obtained from two-dimensional projections onto the imaging plane. For example:

- If an object is spherical, then its volume V can be estimated from the area A of its projection, as

$$V = \frac{4A^{3/2}}{3\sqrt{\pi}}.$$

  The volume could alternatively be estimated from the perimeter or diameter of the projected circular shape, but this is not recommended because these statistics are more variable than the area. In fact, the perimeters and diameters of circular regions are more precisely estimated as functions of the area than directly.

- If an object is a prolate ellipsoid, i.e. the two minor axes are of equal length, and the projection is along one of the minor axes, then the volume V can be estimated from the length L and breadth B of the projected ellipse, as

$$V = \frac{1}{6}\pi LB^2.$$

6.2   MEASURES OF SHAPE

Shape information is what remains once

- location,
- orientation
- and size

features of an object have been dealt with. Therefore two objects that are the same, except that they are in different positions in the image and orientated at different angles, and that one is bigger than the other, will have the same shape. In the computer vision literature, the term *pose* is often used to refer to location, orientation and size. (Of course, invariance will hold only approximately in the case of a lattice representation of an object.)

We shall begin our consideration of shape by returning to the very first example presented in this book, that of the turbinate image (§1.1.1). The objective of image analysis in this application was to estimate a *morphometric index*, which was defined as the ratio of air space area in the cross-section of the nasal cavities to air space area plus turbinate bone area. The index is pose-invariant, and so is a measure of shape. Figure 6.7 shows the turbinate image after the application of a morphological closing to reduce noise levels and improve connectivity, as shown earlier in Fig. 5.6(b). The turbinate bone in each nasal cavity has been isolated from the remaining bone areas using a mouse to control a screen cursor. Numbers showing the areas of the largest

Fig. 6.7 Turbinate image after morphological filtering and manual separation of the turbinate bone in each nasal cavity from the remaining bone areas, together with numbers showing the areas of the largest regions.

black and white regions have been included on the figure. From these areas, the morphometric index of the left nasal cavity can be estimated as

$$\frac{27\ 100 + 1900 + 2100}{27\ 100 + 1900 + 2100 + 11\ 000} = 0.74.$$

Similarly, for the right nasal cavity, the index is

$$\frac{26\ 000 + 2500 + 800}{26\ 000 + 2500 + 800 + 10\ 800} = 0.73.$$

These indices are typical of a pig that is free of the disease atrophic rhinitis. Note that these statistics are dimensionless, as are all measures of shape, because they are all size-invariant.

Returning to generalities, there are a very large number of possible ways of describing shape. However, a few, based on the statistics in § 6.1, are widely used. They will be presented here, except for those based on boundaries, consideration of which will be deferred to §6.3.

Probably the most commonly used shape statistic is a measure of *compactness*, which is defined to be the ratio of the area of an object to the area of a circle with the same perimeter. A circle is used as the basic shape with which to make comparisons because it is the object with the most compact shape (although Rosenfeld (1974) shows that for objects approximated by pixels, an octagon minimizes the measure). The statistic is

$$\text{compactness} = 4\pi \frac{\text{area}}{(\text{perimeter})^2}.$$

Note that this is a shape statistic because it is pose-invariant: changing an object's location or orientation or doubling its size, say, will leave the measure unchanged, except for the proviso (already commented on) that the estimated perimeter of an object can be sensitive to the resolution at which it is measured. The measure takes its largest value of 1 for a circle. Any departures in the object shape from a circular disc, such as an elliptical shape or a border that is irregular rather than smooth, will decrease the measure.

The responsiveness of the measure of compactness to any departure from a circular disc can sometimes be a disadvantage. In some situations, it is useful to have measures that are sensitive only to departures of a certain type from circularity. Such statistics can be devised. For example, a measure of *convexity* can be obtained by forming the ratio of the perimeter of an object's convex hull to the perimeter of the object itself; that is,

$$\text{convexity} = \frac{\text{convex perimeter}}{\text{perimeter}}.$$

This will take the value of 1 for a convex object, and will be less than 1 if the object is not convex, such as one having an irregular border. Also, a measure of *roundness*, excluding these local irregularities, can be obtained as the ratio of the area of an object to the area of a circle with the same convex perimeter; that is,

$$\text{roundness} = 4\pi \frac{\text{area}}{(\text{convex perimeter})^2}.$$

Again, as with the measure of compactness, this statistic equals 1 for a circular object and is less than 1 for an object that departs from circularity, except that it is relatively insensitive to irregular borders. The effects on the measure are more complex if an object is non-convex in an overall sense, for example a horseshoe shape.

Another statistic often used to describe shape is a measure of *elongation*. This can be defined in many ways, one of which is obtained by taking the ratio of an object's length to its breadth:

$$\text{elongation} = \frac{\text{length}}{\text{breadth}}.$$

This statistic was derived for the sheep's eye muscles at the beginning of the chapter. Another measure of elongation is given by the ratio of the second-order moments of the object along its major and minor axes:

$$\frac{\lambda_1}{\lambda_2}.$$

The results are often very similar to the previous measure of elongation.

Another use of moments as measures of shape is that the central moments μ'_{kl} can be made scale-invariant by dividing them by $\mu_{00}^{(k+l+2)/2}$. This converts rotationally invariant moments into measures of shape. Thus, the first *scale and rotationally invariant moment* is

$$\frac{\mu'_{20} + \mu'_{02}}{\mu_{00}^2}.$$

This is another measure of compactness. It measures how dispersed the pixels in an object are from their centroid, in comparison with the most compact arrangement of the pixels. It takes its smallest value of $1/2\pi = 0.159$ for a circular object.

Table 6.3 gives some of these shape statistics, and the size statistics from which they were derived, for the 13 algal cells in Fig. 1.12(f). Measures involving the convex perimeter have been omitted, because all the cells are

Table 6.3 Size and shape statistics for the algal cells in Fig. 1.12(f)

Cell	Area	Perimeter	Length	Breadth	Compact-ness	Elong-ation	$\dfrac{\mu'_{20} + \mu'_{02}}{\mu_{00}^2}$
1	358	68	23	19	0.96	1.23	0.165
2	547	84	28	24	0.97	1.20	0.163
3	294	65	24	15	0.88	1.62	0.182
4	774	100	34	29	0.97	1.15	0.161
5	566	86	29	25	0.96	1.16	0.161
6	306	66	24	14	0.87	1.65	0.183
7	243	57	19	16	0.94	1.21	0.164
8	312	62	20	19	1.01	1.03	0.159
9	277	60	20	17	0.97	1.17	0.163
10	418	78	26	21	0.87	1.28	0.171
11	387	73	26	19	0.92	1.38	0.170
12	261	59	19	18	0.95	1.07	0.161
13	440	77	27	20	0.93	1.31	0.168

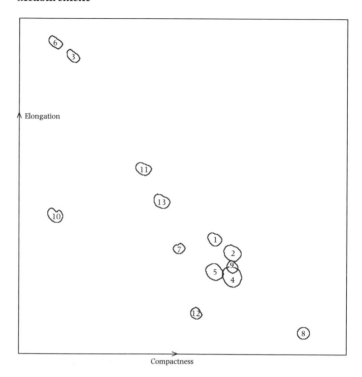

Fig. 6.8   A scatter plot of a measure of elongation against a measure of compactness for the segmented algal cells: each point is represented by that cell's outline.

nearly convex. For these cells, the measure of elongation and $(\mu'_{20} + \mu'_{02})/\mu^2_{00}$ are highly correlated. Figure 6.8 shows elongation plotted against compactness, with each point represented by that cell's outline. Although all the cells have very similar shapes, some differences show up in the figure. For example, cell 8 is the one closest to being circular, and it has the lowest measure of elongation and the highest measure of compactness. Cells 3, 6 and 10 are the least compact, because their perimeters are proportionally longer than those of the other cells. In the case of cells 3 and 6, this is because they are elongated, and they are therefore located in the top-left corner of the figure. The segmented version of cell 10 has an indentation (although probably the cell itself does not), and this is why it is less compact than the others. Because the cell is no more elongated than any of the others, it is located towards the centre-left in the figure.

The description of shape is an open-ended task, because there are potentially so many aspects to an object even after location, orientation and size effects have been removed. Other approaches include the use of *landmarks* (identifiable points on objects) (Goodall, 1991) and warpings such as *thin-plate splines* and other *morphometric methods* (Bookstein, 1991), which consider image plane distortions needed to move landmarks to designated locations. Further methods are discussed in the reviews of shape analysis by Pavlidis (1978, 1980) and Marshall (1989), such as the use of *skeletons* (§5.3) and of distributions of *chord-length*; chords being the lines joining any pair of points on the boundary of an object.

Allometry is a subject related to shape. It is often found that the sizes of different parts of organisms are linearly related on a logarithmic scale. For example, the lengths L and breadths B of members of a population may, on average, obey

$$\log L = \alpha + \beta \log B$$

for certain values of the constants α and β (e.g., Causton and Venus, 1981, Ch. 6). This relationship can be re-expressed as

$$L = e^{\alpha} B^{\beta}.$$

Therefore L/B^{β} is a size-invariant constant for the population, which it may be used to characterize in preference to the shape statistic L/B. To illustrate, Glasbey, McRae and Fleming (1988) found that the log-lengths and log-breadths of potato tubers were normally distributed, with standard deviations of 0.0766 and 0.0646 respectively. The ratio of these standard deviations gives an estimate of β of 1.19. Therefore the allometric constant ratio is $L/B^{1.19}$, which is consistent with the observation that smaller potatoes tend to be more spherical than larger ones.

6.3   BOUNDARY STATISTICS

The first step in studying the boundary of an object is to extract the boundary pixels from the image. We describe an algorithm, first in words and then mathematically, for generating an ordered list of pixels round the 4-connected boundary.

We use d to denote the four possible directions we can take moving between pixels. In keeping with orientations used elsewhere in this book, direction 0 is defined to be across a row, so that from pixel location (i, j) we should move to $(i, j + 1)$. Proceeding clockwise, directions 1, 2 and 3 will be used to denote changes from (i, j) to $(i + 1, j)$, $(i, j - 1)$ and $(i - 1, j)$ respectively. This code is illustrated in the inset of Fig. 6.9. We shall also use the notation $\Delta(d)$ to denote the change in pixel label produced by direction d, so that $\Delta(0) = (0, +1)$, etc.

We shall track round the boundary from pixel to neighbouring pixel in an *anticlockwise* direction. This choice of direction is arbitrary, but is the common

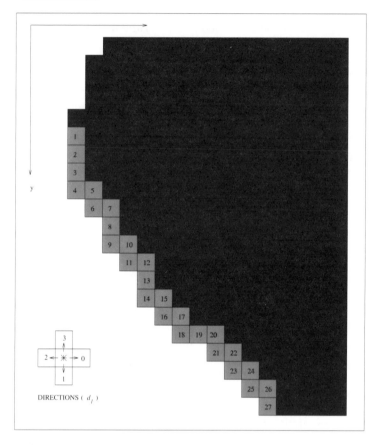

Fig. 6.9 The nose of the fish image, after segmentation: a diagram to illustrate the extraction of a boundary chain-code. The inset shows how directions d_l used in the chain code are defined.

convention. To start, we need two points on the boundary, (y_1, x_1) and (y_2, x_2), where (y_2, x_2) is the next pixel location round the boundary in an anticlockwise direction after (y_1, x_1). Set d_1 to the direction from (y_1, x_1) to (y_2, x_2), and take the initial direction of search (d_2) for (y_3, x_3) to be $(d_1 + 1)$, except that if $d_1 = 3$, we choose $d_2 = 0$.

The algorithm then proceeds as follows. If $(y_2, x_2) + \Delta(d_2)$ is a pixel location in the object then (y_3, x_3) is set to this pixel. Otherwise, d_2 is decreased by 1, except that when it reaches -1 it is reset to 3, and $(y_2, x_2) + \Delta(d_2)$ is examined to see if it is an object pixel. After trying at most four directions, point (y_3, x_3) will have been found, in a direction we preserve as d_2. The search for (y_4, x_4) then begins in direction $d_2 + 1$. Figure 6.9 shows the start of the boundary round the fish (Fig. 1.9e), starting from the nose, and Table 6.4 shows the stored locations. (The fish image was segmented using a semi-

Table 6.4 First few boundary points and chain code
for segmented fish image.

l	y_l	x_l	d_l
1	89	30	1
2	90	30	1
3	91	30	1
4	92	30	0
5	92	31	1
6	93	31	0
7	93	32	1
8	94	32	1
9	95	32	0
10	95	33	1
11	96	33	0
12	96	34	1
13	97	34	1
14	98	34	0
15	98	35	1
16	99	35	0
17	99	36	1
18	100	36	0
19	100	37	1
20	100	38	1
21	101	38	0
22	101	39	1
23	102	39	0
24	102	40	1
25	103	40	0
26	103	41	1
27	104	41	1
.	.	.	.
.	.	.	.
.	.	.	.

automatic method involving thresholding and the use of a computer mouse to
complete the segmentation, rather than by making use of a second, back-illuminated image discussed in §1.2.)

The boundary tracking is complete once we return to the starting point,
except for the important proviso that if the object is only one pixel wide at
(y_1, x_1) then we shall return to this point more than once before having
tracked round the whole boundary. To avoid this case, it is simply necessary
to check that after reaching (y_1, x_1), we move to (y_2, x_2), and then we know
that the algorithm has finished.

The algorithm is formalized as follows:

1. Initialize by setting (y_1, x_1) to a boundary point on the object (set A) and d_1
   to an initial direction such that $(y_2, x_2) = (y_1, x_1) + \Delta(d_1)$ is the next pixel
   location round the boundary in an anticlockwise direction. Also set $l = 2$.

2. Set the search direction $d_l = d_{l-1} + 1$ (mod 4), i.e. an integer between 0 and 3, obtained as the remainder after dividing $(d_{l-1} + 1)$ by 4.
3. $(y_{l+1}, x_{l+1}) = (y_l, x_l) + \Delta(d_l)$.
    If $(y_{l+1}, x_{l+1}) \in A$ then go to step 4;
    otherwise $d_l \rightarrow d_l - 1$ (mod 4), and repeat step 3.
4. If $(y_l, x_l) = (y_1, x_1)$ and $(y_{l+1}, x_{l+1}) = (y_2, x_2)$ then set $N = l - 1$, and the algorithm is completed;
    otherwise $l \rightarrow l + 1$ and go to step 2.

The algorithm can be modified to track an 8-connected boundary, by redefining d and Δ so that 8 directions are considered in anticlockwise order, including the diagonal neighbours.

The *chain code* (Freeman, 1974), consisting of the starting location and a list of directions $d_1, d_2, \ldots, d_N$, provides a compact representation of all the information in the boundary. It can also be used directly to derive the summary statistics of §6.1. For example, in simple cases, N is equal to the statistic N_4 (§ 6.1.2) and can be used to estimate the perimeter. The area and moments of an object can also be obtained from its chain code. But, if an object contains *holes*, the two sets of measures will be different. For example, the perimeter defined in §6.1.2 includes the boundary lengths of holes in an object, whereas N is a measure only of the length of a single connected boundary.

The directions d_l are estimates of the slope of the boundary, but because they are all multiples of $90°$, they are very imprecise. Smoothed estimates may be obtained by considering more distant boundary points. The *k-slope* of the boundary at (y_l, x_l) can be estimated from the slope of the line joining $(y_{l-k/2}, x_{l-k/2})$ and $(y_{l+k/2}, x_{l+k/2})$, for some small, even value of k. This works out as an angle of

$$\tan^{-1}\left(\frac{y_{l+k/2} - y_{l-k/2}}{x_{l+k/2} - x_{l-k/2}}\right),$$

measured in a clockwise direction, with a horizontal slope taken to be the zero. Here, as in §3.2.1, '$\tan^{-1}$' is assumed to produce output over a range of angles of 2π radians.

In a similar fashion, the *k-curvature* of the boundary at point (y_l, x_l) can be estimated from the change in the *k-slope*:

$$\left\{\tan^{-1}\left(\frac{y_{l+k} - y_l}{x_{l+k} - x_l}\right) - \tan^{-1}\left(\frac{y_l - y_{l-k}}{x_l - x_{l-k}}\right)\right\} \quad (\text{mod } 2\pi).$$

Figure 6.1(c), which has already been discussed, shows the curvature of the X-ray eye-muscle boundaries, evaluated with $k = 10$.

Fourier descriptors

Use of *Fourier descriptors* is another approach to describing the boundary of an object. Granlund (1972) proposed approximating $y_1, \ldots, y_N$ by a sum of $K \ (\leq \frac{1}{2}N)$ sine and cosine terms:

$$\hat{y}_l = \bar{y} + \sum_{k=1}^{K} \left\{ a_k \cos\left(\frac{2\pi k l}{N}\right) + b_k \sin\left(\frac{2\pi k l}{N}\right) \right\} \quad \text{for } l = 1, \ldots, N,$$

where

$$a_k = \frac{2}{N} \sum_{l=1}^{N} y_l \cos\left(\frac{2\pi k l}{N}\right),$$

$$b_k = \frac{2}{N} \sum_{l=1}^{N} y_l \sin\left(\frac{2\pi k l}{N}\right) \quad \text{for } k = 1, \ldots, K.$$

And similarly for x, but with different Fourier coefficients. If $K = \frac{1}{2}N$ then the curve passes through all the boundary pixels, whereas if $K < \frac{1}{2}N$ then the curve provides a smooth approximation.

The Fourier coefficients (*a*s and *b*s) can be obtained by regressing y on the sine and cosine terms. The sum of squares of the ys about their mean $\bar{y}$ can be represented as a sum of squares of the Fourier coefficients plus a sum of squares of the differences between the ys and their Fourier approximation:

$$\sum_{l=1}^{N} (y_l - \bar{y})^2 \quad = \quad \frac{1}{2} N \sum_{k=1}^{L} (a_k^2 + b_k^2) \quad + \quad \sum_{l=1}^{N} (y_l - \hat{y}_l)^2.$$

Alternatively, fast Fourier transform algorithms can be used to compute $a_1, \ldots, a_{N/2}$ and $b_1, \ldots, b_{N/2}$, provided that N is either a power of 2 or a product of small prime factors. If necessary, extra boundary points can be introduced by duplicating, or interpolating between, pixels to produce such a value for N. (See §3.2 for more details about the Fourier transform.)

For the segmented fish, the 4-connected boundary has length $N = 1416$. The Fourier approximations using one coefficient only, are

$$\hat{y}_l = 99.4 - 2.6 \cos\left(\frac{2\pi l}{N}\right) + 65.3 \sin\left(\frac{2\pi l}{N}\right)$$

$$\hat{x}_l = 256.4 - 188.5 \cos\left(\frac{2\pi l}{N}\right) + 8.5 \sin\left(\frac{2\pi l}{N}\right) \quad \text{for } l = 1, \ldots, N.$$

From a total sum of squares of the ys of 3 292 000, the residual sum of squares is reduced to

$$3\ 292\ 000 - \frac{1416}{2}(2.6^2 + 65.3^2) = 268\ 000.$$

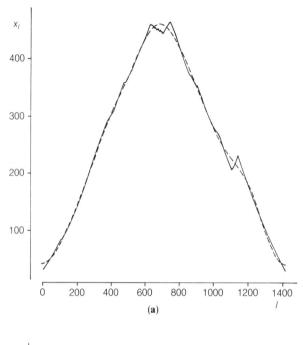

(a)

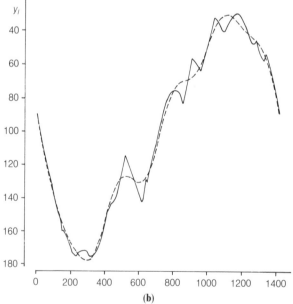

(b)

Fig. 6.10 Boundary points of segmented fish image, and their approximation using $K = 5$ Fourier coefficients, plotted against pixel count l around the boundary, starting from the nose: (a) column coordinate x_l; (b) row coordinate y_l.

Therefore 92% of the variation in the ys has been accounted for. Figure 6.10 shows the fitted curves when $K = 5$, for which 99.2% of the y variability and 99.8% of the x variability has been explained. Figure 6.11 shows the resulting boundaries when $K = 1$, 5 and 20. When $K = 1$, the result is an ellipse. Increasing the number of terms improves the fit, but at the price of using

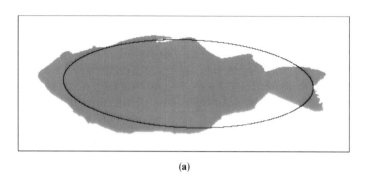

(a)

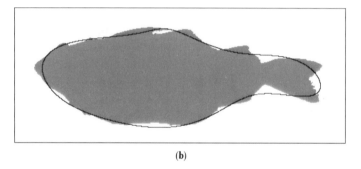

(b)

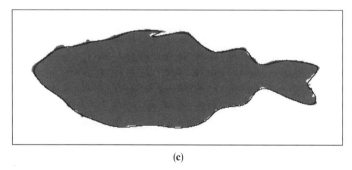

(c)

Fig. 6.11 Segmented fish image, together with approximations to the boundary based on Fourier descriptors using different numbers of coefficients K: (a) $K = 1$; (b) $K = 5$; (c) $K = 20$.

more parameters. For $K = 20$, 99.97% and 99.99% of the variation in y and x respectively have been accounted for.

Crimmins (1982) shows how the Fourier coefficients can be transformed to provide measures of shape, i.e. statistics that are invariant to changes in an object's pose. There are other ways of using Fourier descriptors to approximate object boundaries. For instance, Zahn and Roskies (1972) proposed a set of Fourier descriptors based on *local boundary slope* rather than directly on the (y, x) coordinates. Rohlf and Archie (1984) and Mou and Stoermer (1992) compared alternative approaches, and applied Zahn and Roskies' method to describe the outlines respectively of mosquito wings and diatoms. It is also possible to use *polygons* (Pavlidis, 1977), *splines* (Hill and Taylor, 1992) and *conic sections* (Bookstein, 1978) to describe boundaries. Marshall (1989) reviews the use of stochastic descriptions of boundaries based on *circular autoregressive models*.

6.4 SUMMARY

The key points of this chapter are as follows:

- Measurements are usually taken from the images output from segmentation algorithms (Chapter 4), which have possibly also been processed using morphological operators (Chapter 5). In some applications, measurements can be obtained directly from the original image.
- Three general categories of measurement are
  — size;
  — shape, independent of size;
  — boundary statistics.
- Two types of size statistics were considered:
  — those based on moments, such as
    - area,
    - centre of gravity or centroid,
    - second-order moments, such as the moment of inertia, which is invariant to changes in location and orientation;
  — those based on distances:
    - average breadth,
    - length of a curved line, such as a boundary,
    - Feret diameters, and convex-hull perimeter.
- Shape statistics that were considered included measures of
  — compactness;
  — convexity;
  — roundness;
  — elongation;
  and scale- and rotationally-invariant moments.

- Boundary descriptors considered were
  - ordered set of boundary pixels;
  - chain code;
  - local measures of slope and curvature;
  - Fourier descriptors.

7

Summary of Results

This chapter summarizes what has been achieved for each of the images used in the book. Sometimes an operation on an image was performed simply to illustrate an idea, but on other occasions progress was made towards the aims described in §1.1.3. The success in doing this varied, and we shall try to make a fair assessment of what has been accomplished. In some cases, more complex techniques, beyond the scope of this book, would be called for, and we shall point towards relevant literature. In other cases, techniques have not been adequately developed at the time of writing: image analysis is an active area of research.

The layout of this chapter will be a series of summaries of the achievements for each of images listed in Table 1.2.

- *Algal cells* The DIC microscope image of algal cells is shown in Fig. 1.6(a). The aim was to count and measure the cells. The first stage was to apply a threshold to the image in order to segment the darker parts of the cells from the background (Fig. 1.12(d); thresholding is discussed in §4.1.1). Then a morphological opening was used to separate pairs of touching cells and reduce the roughness of the cell boundaries (Fig. 1.12(e); discussed in §5.2). Alternatively, the watershed algorithm (§4.3) could have been applied to the distance transform of the binary image (§5.2), as described by Serra (1982, pp. 413–416).

  Size and shape statistics of cells were extracted from the segmented image. These included moment statistics (Table 6.1), perimeters, lengths and widths, and measurements of shape such as compactness and elongation (Table 6.2). Figure 6.8 is a display of the relative shapes of the dark parts of the cells. The statistics could be used to discriminate between different types of algae.

  The measurement is successful, although most of the statistics have been affected by the fact that the bright parts of the cells were lost after the thresholding. This problem could be overcome by calibration, or by an approach based on multiple thresholds and morphological operations. More sophisticated approaches could take account of the specific optics of DIC microscopy (Holmes and Levy, 1987, 1988).

- *Cashmere fibres*   The microscope image of cashmere goat fibres is shown in Fig. 1.6(d). The image was obtained in order to measure the fibre diameters. This is made more difficult because the fibres appear differently according to their position relative to the focal plane. The image was used in Fig. 3.1 to illustrate the different types of linear filter (smoothing, edge finding and unsharp masking). The amplitude and phase of the Fourier transform of the image are shown in Figs 3.5(c) and (d).

  In §5.5, a sequence of greyscale morphological operations was used to determine the distribution of diameters. There is, however, scope for further refinement of this approach, and for taking account of bias in diameter estimates of fibres at different focal planes—see, for example, Glasbey *et al.* (1994).

- *DNA sequencing gel*   The autoradiograph is shown in Fig. 1.9(c). This example of an electrophoretic gel is one-dimensional because gene fragments only migrate in one direction—down the separate tracks. In Fig. 1.10, we showed the difference between the original photographic plate and the digitized image. The positions of the bands in each track are of interest. We did not attempt to automatically find the band positions, but used some techniques to make this easier for the human interpreter.

  In §3.2.3, we showed how it is possible to reverse blurring in an image, by deconvolution using the Wiener filter. This involved modelling the degradation process, in order to identify the optimal linear filter. In §3.4.2, we showed how Prewitt's edge filter could be used to find edge directions, the results of which are displayed in Fig. 3.18. These orientations can be used to remove warping from an autoradiograph (Glasbey and Wright, 1994). In §5.4, the image was used to illustrate the measurement of texture based on the auto-crossproduct function, and in §5.5 it was shown how the top-hat transform could be used to subtract the trend in background brightness and so make the bands more distinct.

  For work on automatic detection of bands in DNA sequencing gels, see Griffin and Griffin (1993), Khurshid and Beck (1993) and Smith (1993). Analysis of one-dimensional gels in general is considered by Trubuil (1993).

- *Electrophoretograms*   Figures 1.9(a, b) show a pair of SDS–PAGE gel electrophoretograms. The aim was to recognize differences between the gels. In §2.5, it was shown how the gels may be aligned using registration techniques. Then the colour composite, shown in Fig. 2.17, makes it easy to identify the differences in location of a few spots. Meyer and Beucher (1990) used greyscale morphological operations to identify spots. It is then possible to estimate protein volumes by summing the pixel values (§6.1.1). Many attempts to automatically analyse 2D electrophoretic gels have been described. See, for example, Appel *et al.* (1991), Conradsen and Pedersen (1992), Nokihara, Morita and Kuriki (1992) and Solomon and Harrington (1993).

- *Fish*   The fish image is shown in Fig. 1.9(e). The aim was to extract information about the shape of the fish, which could then be used, in conjunction with images of many other fish, to discriminate between different species. There are many ways of quantifying shape. Strachan *et al.* (1990a) looked at statistics based on dividing the fish into a number of quadrilaterals and measuring their orientation and average grey level. Glasbey *et al.* (1994) explored two statistical theories of shape, those of Procrustes methods (Goodall, 1991) and morphometrics (Bookstein, 1991).

  In §2.3.2, the fish image was used to illustrate the way that pseudocolour makes greyscale comparisons easier than in a standard display. In §6.3, we showed how the fish outline could be described using a boundary chain code, once the fish has been distinguished from its background. The segmentation was achieved using a semi-automatic method involving thresholding and the use of a computer mouse to complete the boundary. An alternative approach, discussed in §1.2, is to make use of a second, back-illuminated image. Figure 6.11 shows the fish boundary approximated by varying numbers of Fourier descriptors. Since these summarize aspects of the shape of the outline, it should be possible to use a selection of the Fourier descriptors to construct a discrimination between species. For more recent work, involving the use of colour in discrimination, see Strachan (1993).

- *Fungal hyphae*   The digitized photograph of part of a fungal mycelium is shown in Fig. 1.9(d). The image was obtained in order to further understand the spatial distribution of hyphae. The image was used to illustrate binary image display in §2.1 and zooming and reduction in §2.4.

  In §5.3, a thinning operation was used to obtain the skeleton. This reduced the hyphae to the thickness of a single pixel, which is convenient for estimating the total length of hyphae. In §5.4, the spatial arrangement of the hyphae was examined by calculating the auto-crossproduct function. This showed no obvious pattern, suggesting a uniform distribution of orientations. The total hyphal length was estimated in §6.1.2, with allowance made for effects due to lines being represented as lattice points rather than in continuous space. Further work on fungal morphology, and its relationship to soil structure, is described by Crawford, Ritz and Young (1993).

- *Landsat*   Thematic Mapper bands 1–5 and 7 from the Landsat satellite are shown in Figs 1.8(a–f) for a region on the east coast of Scotland. In combination, they form a multispectral image. There are many uses for such images, ranging from cartography to environmental monitoring. A comprehensive reference is Colwell (1983).

  In §2.2.2, it was demonstrated how the greyscales of individual bands could be stretched to make features easier to recognize. In §2.3.3, a natural colour display was produced from bands 1, 2 and 3. This is shown in Fig. 2.10. A standard use of satellite images is to classify ground cover. This can be done using the multivariate and contextual classification methods described in §4.1.2 and 4.1.3. Facilities to do this are available in most software designed for remotely sensed image handling. In §6.1.1, we showed how areas of a particular crop type (oil-seed rape) could be estimated after multivariate classification.

- *Magnetic resonance images*   MRI inversion recovery and proton density images are shown in Figs 1.7(a) and (b). Together, they constitute a bivariate multimodal image. The interest was in estimating the cross-sectional areas of different tissue types. In §6.1.1, we briefly discussed the thorough job done by Fowler *et al.* (1990) in calibrating their experiment, to estimate volumes and water content.

  Figure 2.11 is a colour display of the bivariate image, which reveals differences between tissue types more readily than does an examination of the two separate images. In §4.1.2, linear discriminant analysis was used to classify pixels into four tissue types: lung tissue, blood in the heart, muscle and subcutaneous fat. In addition to this, the images were used to illustrate the possibilities for unsupervised classification using k-means clustering.

  Because MRI imaging is widely used, much work has been done on the automatic analysis of such images. See, for example, Colchester and Hawkes (1991) and Barrett and Gmitro (1993). Recent interest has centred on analysis of three-dimensional images and on multimodality— the synthesis of images produced by different technologies.

- *Muscle fibres*   Figure 1.6(b) is a stained section through a muscle, showing three fibre types. The aim was to count and measure the sizes of fibres of the different types.

  This image was first used, in §2.2.2, to illustrate histogram equalization. This made the variation in pixel values within each fibre easier to see (Fig. 2.7d). The muscle fibres image was used extensively in Chapter 4 to illustrate different approaches to segmentation, including thresholding (§4.1), edge-based methods (§4.2) and a region-based method using the watershed algorithm (§4.3). Fibre sizes in a segmented image can be measured using the methods described in Chapter 6.

  It has proved difficult to achieve a fully automatic segmentation of this image, and work is still in progress. The main difficulty is that some divisions between fibres are very faint. The human eye, in recognizing fibres, is making use of the fact that the fibres seem to have a convex shape and a narrow range of sizes. It seems likely that any fully automatic segmentation would also need to make use of this. More-complex segmentation methods are discussed at the end of Chapter 4.

- *SAR*   A synthetic aperture radar image is shown in Fig. 1.8(g). This image can be used in the same way as the Landsat data, for application in areas such as cartography and environmental monitoring, but the interpretation is more difficult because of the presence of speckle noise. Also, the way radar interacts with vegetation and the ground is not yet fully understood. However, the properties of the speckle noise are known, which helps in constructing suitable processing methods.

In §2.2.2, the log transform was introduced, which ensures that the speckle variance is constant throughout the image. In Chapter 3, a variety of smoothing filters were tried, including the moving median (§3.3.1) and spatially-adaptive filters (§3.3.2). Filters that make use of our knowledge of the speckle variance were also considered: the Wiener filter (§3.2.3) and Lee's filter (§3.3.2).

In Chapter 4, three ways were explored for segmenting the image into separate fields. In §4.1.3, contextual classification was used to segment the image into bright and dark fields using post-classification smoothing (Fig. 4.9b) and an ICM algorithm (Fig. 4.9c). In §4.3, a split-and-merge algorithm divided the image into regions for which the variance of pixel values was less than a specified limit. For other approaches to smoothing and segmenting SAR images, see Durand *et al.* (1987) and Oliver (1991).

- *Soil aggregate* Figure 1.6(c) shows a cross-section of a soil aggregate, imaged by electron microscopy. It was obtained in order to study the properties of the pores in the soil material.

  In §4.1, manual and automatic thresholding methods were used to recognize which pixels belong to the pore space. The result is quite rough, so a morphological closing was used to reduce the noise (Fig. 5.1b). In §5.2, a distance transform was used to measure an important property of pore space, namely the distribution of distance from the soil material. The use to which this can be put in modelling pore structure is beyond the scope of this book: see Glasbey *et al.* (1991) and Horgan and Ball (1994). For other recent work on pore structure, which examines its fractal nature, see Bartoli *et al.* (1991) and Young and Crawford (1991).

- *Turbinate bones* This image is shown in Fig. 1.2(a). A manually enhanced version, using black ink and typists' correction fluid, is shown in Fig. 1.2(b). The aim was to measure the ratio of air-space area in the cross-section of the nasal cavities to air-space area plus turbinate-bone area.

  In §5.2, a morphological closing was used to remove much of the irrelevant speckle that had resulted from the 'printing' described in §1.1.1. Manual intervention was required to separate the turbinate bones from the remaining bone areas, after which calculation of the morphometric index was straightforward (§6.2). Although a fully automatic algorithm would be desirable, what we describe here is a big improvement over the approach originally used.

- *Ultrasound* An ultrasound image of a cross-section through a sheep's back is shown in Fig. 1.7(d). The aim was to estimate depths of fat and muscle tissue. The features of this image proved too irregular, and too much affected by irrelevant artefacts of the image capture to allow a successful segmentation or classification based on the methods described in this book.

A manual segmentation was required, after which the measurement of properties of the image is straightforward (§6.1.2). More complex manipulation of the pixel values is needed, using methods akin to the Hough transform, to find the boundaries automatically—see Glasbey, Abdalla and Simm (1994).

- *X-ray CT*   Figure 1.7(c) shows an X-ray computed tomograph of a sheep. Tomographic reconstruction of images from projections is not covered in this book (see e.g. Rosenfeld and Kak, 1982, Ch. 8). The reason for collecting the image was to estimate areas of fat and lean tissue.

In §2.2.2, a piecewise linear transform was used to make it easier to see differences in tissue type. The image was used extensively in Chapter 3 to illustrate different linear filters (§3.1) and nonlinear smoothing (§3.3) and edge-detection (§3.4) filters, and to demonstrate the information content of the amplitude and phase of the Fourier transform (§3.2.1), as well as the effects of constructing filters in the frequency domain (§3.2.3).

After thresholding, a morphological opening was used in §5.2 to obtain a smooth outline for the eye muscles in the image, which were then measured in §6.1.2. It was also shown how Feret diameters and shape statistics (§6.2) could be obtained. The boundary curvature was estimated (Fig. 6.1c) and the concept of a convex hull illustrated (Fig. 6.6). In general, similar algorithms are applicable to MRI and to X-ray CT images.

The above discussion illustrates a wide variety of aims and solutions. Techniques have been drawn from the different stages of image analysis—*Display, Filtering, Segmentation, Mathematical Morphology* and *Measurement*—as appropriate. The success achieved using the methods covered in the book has varied. For example, the aims have been completed with the fungal hyphae and algal images; some more refinement is needed with the SAR and muscle fibres images; and considerably more work is needed with the ultrasound and electrophoresis images.

The material covered in this book will not enable our readers to solve all the image analysis problems they may encounter. Rather, we hope that it lays a foundation of the main ideas on which the subject is built. If the reader wishes to go further, there are many excellent, more advanced books and journals that can be consulted. Their content should be more comprehensible after study of our book. The literature continues to increase, and we expect image analysis techniques to continue to grow in power and utility.

Appendix

Computer Hardware and Software

In this appendix we discuss some of the issues in choosing hardware and software for image analysis. The purpose is to draw attention to the issues involved rather than to discuss the merits of particular manufacturers' products. This latter information would date very rapidly. The reader should consult recent computer journals and magazines, particularly those devoted to graphics. Even the issues involved may change with time.

Changing computer technology also affects the amount of time required by image analysis algorithms. The computing power available for a given amount of money appears to be increasing exponentially, and has been estimated as doubling every two years. The complexity of commercially available software has been growing at a similar rate.

In the two following sections, we separately consider hardware and software requirements. Sometimes, they will be offered for sale as a package.

A.1 HARDWARE

Hardware consists of the physical components of an image analysis system. The main components are illustrated in Fig. A.1.

It will be essential to have the following:

- A *central processing unit (CPU)*. This resides on a chip, and does the computing. It will have some *memory* to store the information it is currently using. There may be a memory specifically for storing digital images, usually called a *framestore*. This is often an added component in a computer system.
- A *display monitor* to look at digital images.
- Some image *storage* device to store digital images.

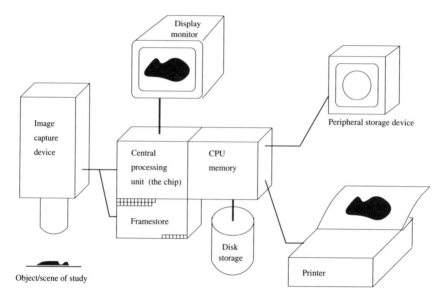

Fig A.1 Basic hardware components of an image analysis computer system

- An *image capture device* to create digital images from whatever is being studied.

It can also be useful to have the following:

- Some *peripheral storage media*. These are ways of storing many images at low cost, although access will be slower. They also provide backup security should images on the main storage device become lost.
- A *printer* or other device to produce permanent copies of images.

There is an increasing trend for computing equipment to be joined to a network, and this applies to most of the equipment mentioned above. This has the advantage of allowing equipment and data to be easily shared among several users.

Since it will be necessary to acquire several items of equipment, an important issue will be *compatibility*—the extent to which different pieces of equipment can work properly together. It is the authors' experience that this can be a substantial problem. Two ways to overcome this are to buy all equipment from one manufacturer, or to buy a packaged system. These options may cost more than a separately bought configuration, and can be less flexible.

We next consider the issues involved in choosing each of the items of equipment mentioned above.

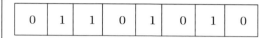

0	1	1	0	1	0	1	0

A byte is the fundamental unit of computer memory.
It consists of 8 bits, each of which may be 0 or 1.
Depending on context, it may encode a character, an integer number
between 0 and 255, or some other information.
For example, the byte above represents the number 106 in binary notation
or the character "j" (in standard ASCII character coding).
More complex information, such as numbers with a wider range, are coded
using several bytes.

Fig A.2   Storing information in bytes

A.1.1   The central processing unit

The central processing unit (or chip) is the 'brain' of the computer. Although
different manufacturers may produce computers with the same chip, the
differences between them will be much less than those between computers
with different chips. The chip is the main factor in determining the speed with
which a system works. Naturally, faster chips tend to cost more. It can be
difficult to judge before a system is in routine use how long it will take to complete particular tasks. If possible, advice should be sought from others working
on similar tasks.

Associated with a CPU chip there will be a certain amount of CPU
memory, which stores the information currently required where it can be
accessed very quickly. All other information storage is much slower. The
basic unit of computer memory is the *byte* (see Fig. A.2), and CPU memory is
usually measured in *megabytes (Mb)* or millions of bytes (actually 2^{20} bytes).
CPU memory is relatively inexpensive, and it is usually worth spending money
to have a good amount of it. The link between the CPU memory and the CPU
(the *bus*) affects memory access, and therefore processing, times. There are
also aspects to how the CPU memory is managed that will affect performance.
For example, a *cache* memory on the CPU allows blocks of data to be processed
and transferred quickly.

Most computers have only one CPU. Computers with more than one
are also built, and designed so that computing tasks can be shared out
and performed on different CPUs. This is *parallel computing*. Many image
analysis tasks involve repeating the same operation many times, and are well
suited to parallel implementation.

A.1.2   The image capture device

The image capture device is the equipment that provides the link between the real world and the digital image in a computer. It can be any machine that records a signal from the object under study and converts it to a set of pixel values. Such devices may be roughly divided into three types: electronic cameras, scanners and other imaging devices.

Electronic cameras are like the familiar video camera, which produces, as an electronic signal, images of the scene it is pointed at. Usually this is an analogue signal, but an *analogue-to-digital converter* will turn it into a digital signal, which may be arranged to form an image. This conversion is often done in the computer by an image framestore or framegrabber. A *still video* camera operates very like a photographic camera, but can store images in digital form, which can later be transferred to a computer.

Scanners work like photocopiers. They are widely available for use in desktop publishing (DTP). Scanners intended for scientific use are also produced. They are designed to produce more precise and repeatable results, and are more expensive. For most purposes, DTP scanners should be adequate. Although the principles they use are similar to cameras, they differ in that

- They can only scan flat objects placed in them, such as photographs. As a result, they are not suitable for freezing an image at a chosen moment while observing a changing scene.
- An ordinary photograph must be produced as an intermediate stage between the scene and the scanner, except in rare cases where objects (such as electrophoresis gels) may be placed directly on the scanner.
- They can often produce higher spatial resolution and more pixels than cameras.
- Hand-held scanners, which are a cheaper option, produce poorer quality images.

Many other types of imaging equipment, particularly medical imaging equipment such as magnetic resonance imagers, ultrasound scanners and positron emission tomographs, produce their images in digital form. From here, they can be transferred to a computer. Sometimes, substantial problems of hardware and software compatibility have to be overcome to do this.

The two critical aspects of any image capture equipment are its spatial and radiometric resolutions. The spatial resolution is basically the number of pixels per unit area the device produces. It determines how much detail is recorded in the object being studied. The radiometric resolution is the number of grey levels recorded. This must be at least two (for a binary image). It is sometimes 16 (half a byte), which is the minimum that can be considered useful in most applications, and is often 256 (a full byte). More than 256 is of little value in many applications.

Another aspect of image capture devices is whether they record colour. As described in chapter 2, this is done by recording more than one value at each pixel. Colour is a useful facility to have in some applications. The question to ask is: Can one see in a colour image features of interest that cannot be seen as clearly in a monochrome image? If so, then colour may help in the image analysis task planned. If not, it may simply add expense and extra processing effort without any benefits.

A.1.3   The framestore

Image analysis software will sometimes store the images it is working with in the CPU memory of the computer. In other cases, it will require a separate memory device in the computer, usually termed an image framestore or graphics device. There are many of these available. Usually they are more than simply storage devices, and can do some elementary processing of the image (such as zooming and arithmetic operations) much more quickly than can be done by the CPU. They may also be the means whereby an image in some analogue form (e.g. a video camera signal) is converted into a digital image.

The size of the framestore will determine how much image data it can hold. The maximum image size will be determined by the number of rows and columns of pixels available in the framestore. The number of grey levels that can be handled is usually expressed in terms of the number of bits (see Fig. A.2). Eight bits can accommodate $2^8 = 256$ grey levels. To store three such images simultaneously, usually in order to handle red, green and blue components, requires 24 bits. Some framestores will have extra bits to enable features to be drawn on an image without changing pixel values. Naturally, bigger framestores cost more.

A.1.4   Image storage

For many applications, it will be necessary to store digital images somewhere, so that they can be accessed on the computer in the future without the need to capture the image again. Computers usually store information on a disk, locally, or perhaps centrally if the computer is part of a network. Images stored here can be accessed quickly. Images for which rapid access is not needed can be stored on some other peripheral device. These include diskettes, cassettes, tapes of various sorts and various optical storage media. With the exception of diskettes, these are cheaper per amount of information stored than computer disks. They are all considered to be more secure than computer disks. Diskettes can store only a little information each, but are easy to use, convenient for transferring information between computers and can allow an individual more control, since they are not a shared facility. Compact disc read-only memory (CD-ROM) devices are an inexpensive way of storing a large amount of data or images, although they can be written to only once.

In deciding what storage is to be used, the amount of storage needed should be considered. Digital images occupy a lot of space. If the images are to be $n \times n$ pixels in size, with b bytes per pixel, then each image will occupy at least bn^2 bytes. For example, a 512×512 pixel image of 1-byte pixels will occupy $\frac{1}{4}$ Mb.

If lots of images are to be stored, consideration should be given to *image compression* techniques. These are ways of storing images in smaller amounts of storage space. They make use of the property that adjacent pixels tend to be similar, or identical, in value, and so image detail can be stored without the need to record each individual pixel. Compression has the disadvantage that images take longer to access. Some image analysis programs will offer compression as an option. Alternatively, stand-alone programs can be used to compress image (and possibly other) files stored on the computer. They can then be uncompressed (re-expanded) before use.

Many compression algorithms for binary images are very straightforward. For example, we can record for each row of the image what the starting value (0 or 1) is, and at which pixel positions along the row the value changes. This information is sufficient to reconstruct the image, and a considerable reduction in storage space can be achieved. With the turbinate image, the storage required using this algorithm is only 7% of the original uncompressed image. Images with less fine detail would be compressed even more.

Image compression can be with or without loss of information, sometimes termed *lossy* or *lossless*, respectively. If information is not lost then the uncompressed image should be identical to what it was before compression. In some situations, some degradation of fine detail can be accepted in the compression process, in order to achieve higher compression ratios (the ratio of the size of the uncompressed to compressed images). Careful comparison of before and after compression images should be made before such algorithms are used.

Compression algorithms, whether with or without loss of information, differ in their speed of processing and the compression ratios they achieve. For a detailed discussion, see Jain (1989, Ch. 11). At the time of writing, a very popular and widely available algorithm for lossless compression is the Lempel–Ziv algorithm (Lempel and Ziv, 1986). Similarly, for compression with loss, the JPEG standard (Wallace, 1991) is widely used. For descriptions of some other recent algorithms, see Devore, Jawerth and Lucier (1992), Nasiopoulos, Ward, and Morse (1991) and Martinelli, Ricotti and Marcone (1993).

A.1.5   Display

It is essential to be able to display the digital image on a monitor of some sort. Many of the issues in image display were discussed in Chapter 2. Display monitors will differ in the number of pixels and the number of colours or levels of grey that can be displayed. Be aware that some systems may not allow all colours to be displayed at the same time, although this may not be as serious a drawback as it might seem—see §2.3.3.

Some framestores will be configured to use a dedicated monitor for image display, separate from the standard monitor on the computer.

A monitor with an adequate number of pixels and number of colours should be chosen. Naturally, colour monitors are more expensive than monochrome, and cost increases for monitors with greater numbers of display pixels. Monitors may have bigger screens without necessarily having more pixels. However, this may make the screen easier to look at

A.1.6   Printing

It may be useful to produce printed copies of images. Most printers can produce some form of image printout, but the quality of the result can be very variable. Printers designed mainly for printing text will tend to produce very poor reproductions of images, only suitable for very limited purposes. Printers designed for desktop publishing, such as laser printers, will produce much better results. However, they usually print by using a very fine matrix of black dots. An image printed in this way will usually look inferior to a computer monitor display, and may not be acceptable for publication unless the image has little fine detail. Printers that can truly handle grey levels are, at the time of writing, rare and expensive, but this may change in the future. Also, colour printers are several times more expensive to buy and run than monochrome printers. The remarks made above about the number of colours that can be displayed on a monitor also apply to printers.

A variety of other devices that can produce permanent copies of images are also available. Slidewriters, for example, print copies of images onto standard 35 mm transparencies. Finally, a crude but quite effective way of getting permanent copies of digital images is to photograph the monitor.

A.2   SOFTWARE

A large amount of image analysis software, with a wide range of abilities, costs and computer customization is now available. Any attempt to review it would rapidly become out of date. We can only give some general pointers in this section to the issues to be considered.

Table A.1. Facilities in image analysis packages. These facilities should not be considered to be requirements, unless they are needed for the tasks the package is to perform.

Image input facilities—how an image can be read into the package.
Image spatial resolution. Is the number of pixels fixed or variable?
Handling of greyscale images.
Handling of multiple/colour images.
Bus and processing speeds—how will these affect efficiency of use?
Image storage—formats available.
Image printing.
Zooming and scrolling—looking at the image in detail.
Contrast stretching/enhancement.
Image registration.
Filters (smoothing, edge-enhancing etc.)
Segmentation algorithms.
Binary morphological operations.
Greyscale morphological operations.
Image feature measurement.
Image editing—manually editing image features or the results of analysis algorithms.
Algorithm building—ability to store a sequence of operations for single-step execution.
Ability to incorporate user-written algorithms in high-level languages.

A.2.1 Choosing software

The first thing to realize is that all computers will be running basic computer management software called the *operating system*. Most are computer-specific, but some operating systems (e.g. Unix and DOS) are available on different computers, and some computers have a choice of operating systems. In some cases, extra management tools (sometimes called *user interfaces*) will be provided as well. Window systems are an example. Many software packages will work with only one operating system, and will sometimes require a particular user interface to be available.

Some image analysis programs will require particular hardware, usually framestores, to be present. They may or may not offer a choice of framestores with which they will work.

The most important things to examine in any image analysis software are the facilities it provides. These should be considered in relation to the image analysis tasks to be performed. If possible, the package should be tested using images of the type to be studied. A list of the facilities that may be found in many packages is given in Table A.1.

Flexibility is an important feature, and one that can be hard to judge before a program has been used extensively. The program may perform its tasks in a very restricted way, or it may allow the user to modify how it does things.

One point to remember is that flexibility is often sacrificed to greater ease of use. In the long term, the gains in ease of use may not justify the loss of flexibility. A well-written package should have both ease of use and flexibility, in such a way that the user can do straightforward operations very readily, and progress to the more advanced features that provide the flexibility required when experience and confidence have been gained.

Many packages provide their image manipulations in the form of building blocks that can be put together by the user to construct whole tasks. This is a form of programming, and can provide great flexibility. It is useful if the program also provides some standard combinations of operations through an easy-to-use interface. It is also sometimes possible to write image analysis software from scratch using standard programming languages such as C or Fortran. This gives the ultimate in flexibility, but requires a lot of time and effort, and should only be attempted by those with experience of these languages. Libraries of image analysis routines in these languages can be bought, and incorporated into one's own programs.

Finally, it is worth noting that in the authors' experience there is much less correlation between cost and such qualities as usefulness, power, flexibility and reliability in software than there is in hardware. We have found that even software that is free of charge and available in the public domain can be powerful and flexible. With more expensive packages, one will usually get more support from the software's producers.

A.2.2 Software compatibility and image formats

Sometimes a user may wish to handle images in more than one program. For example, an image analysis program can produce scientific results, and a DTP program can be used to incorporate images into documents. Also, one may wish to use more than one image analysis package, perhaps on different computers. This gives rise to the question of compatibility. It arises because digital images may be stored in computer files in several different formats. These differ in how the pixel values are arranged, what extra information is included in the file, etc. At the time of writing, there is a plethora of different formats in use (TIFF, GIF, RLE, PPM, BMP, Sunraster, PCX and many, many more). Most packages will read and write a subset of these, and compatibility will be a problem if these subsets do not overlap. Standardization may come in the future, but until then the user will need to know in what format their images are being stored, and what other possibilities are available in the programs being used. If necessary, programs (many of which are in the public domain) that convert between different formats may be used.

References

Adams, R. (1993). Radial decomposition of discs and spheres. *CVGIP: Graphical Models and Image Processing*, **55**, 325–332.

Adams, R. and Bischof, L. (1994). Seeded region growing. *IEEE Transactions on Pattern Analysis and Machine Intelligence*, **16**, 641–646.

Amit, Y., Grenander, U. and Piccioni, M. (1991). Structural image restoration through deformable templates. *Journal of the American Statistical Association*, **86**, 376–387.

Appel, R. D., Hochstrasser, D. F., Funk, M., Vargas, J. R., Pellegrini, C., Muller, A. F. and Scherrer, J. R. (1991). The MELANIE project—from a biopsy to automatic protein map interpretation by computer. *Electrophoresis*, **12**, 722–735.

Arcelli, C. and Sanniti di Baja, G. (1989). A one-pass two-operation process to detect the skeletal pixels on the 4-distance transform. *IEEE Transactions on Pattern Analysis and Machine Intelligence*, **11**, 411–414.

Asano, T. and Yokoya, N. (1981). Image segmentation scheme for low level computer vision. *Pattern Recognition*, **14**, 267–273.

Aykroyd, R. G. and Green, P. J. (1991). Global and local priors, and the location of lesions using gamma-camera imagery. *Philosophical Transactions of the Royal Society, London, Series A*, **337**, 323–342.

Banfield, J. D. and Raftery, A. E. (1992). Ice-floe identification in satellite images using mathematical morphology and clustering about principal curves. *Journal of the American Statistical Association*, **87**, 7–16.

Barrett, H. H. and Gmitro, A. F. (eds) (1993). *Information Processing in Medical Imaging. Proceedings of the 13th International Conference on Information Processing in Medical Imaging.* Springer-Verlag, Berlin.

Bartoli, F., Philippy, R., Doirisse, M., Niquet, S. and Dubuit, M. (1991). Structure and self-similarity in silty and sandy soils—the fractal approach. *Journal of Soil Science*, **42**, 167–185.

Bernstein, R. (1976). Digital image processing of Earth observation sensor data. *IBM Journal of Research and Development*, **20**, 40–56.

Besag, J. (1986). On the statistical analysis of dirty pictures (with discussion). *Journal of the Royal Statistical Society, Series B*, **48**, 259–302.

Bleau, A., Deguise, J. and Leblanc A. R. (1992). A new set of fast algorithms for mathematical morphology (2): identification of topographic features on grayscale images. *CVGIP: Image Understanding*, **56**, 210–229.

Blum, H. (1973). Biological shape and visual science. *Journal of Theoretical Biology*, **38**, 205–287.

Bookstein, F. L. (1978). *The Measurement of Biological Shape and Shape Change.* Springer-Verlag, Berlin.

Bookstein, F. L. (1989). Principal Warps: thin plate splines and the decomposition of deformations. *IEEE Transactions on Pattern Analysis and Machine Intelligence*, **11**, 567–585.

Bookstein, F. L. (1991). *Morphometric Tools for Landmark Data: Geometry and Biology.* Cambridge University Press, Cambridge.

Borgefors, G. (1986). Distance transforms in digital images. *Computer Vision, Graphics and Image Processing,* **34**, 344–371.

Canny, J. (1986). A computational approach to edge detection. *IEEE Transactions on Pattern Analysis and Machine Intelligence,* **6**, 679–698.

Causton, D. R. and Venus, J. C. (1981). *The Biometry of Plant Growth.* Edward Arnold, London.

Chatfield, C. (1989). *The Analysis of Time Series: An Introduction.* 4th edition. Chapman and Hall, London.

Chen, Y. S. and Hsu, W. H. (1990). A comparison of some one-pass parallel thinnings. *Pattern Recognition Letters,* **11**, 35–41.

Chin, R. T. and Yeh, C. (1983). Quantitative evaluation of some edge-preserving noise-smoothing techniques. *Computer Vision, Graphics and Image Processing,* **23**, 67–91.

Cho, S., Haralick, R. and Yi, S. (1989). Improvement of Kittler and Illingworth's minimum error thresholding. *Pattern Recognition,* **22**, 609–617.

Chow, C.K. and Kaneko, T. (1972). Automatic boundary detection of the left ventricle from cineangiograms. *Computers and Biomedical Research,* **5**, 388–410.

Cochran, W. G. (1977). *Sampling Techniques.* 3rd edition. Wiley, New York.

Colchester, A. C. F. and Hawkes, D. J. (eds) (1991). *Information Processing in Medical Imaging. Proceedings of the 12th International Conference on Information Processing in Medical Imaging.* Springer-Verlag, Berlin.

Colwell, R.N. (ed.) (1983). *Manual of Remote Sensing,* 2nd edition, Volume 2. American Society of Photogrammetry, Falls Church, Virginia.

Conradsen, K. and Pedersen, J. (1992). Analysis of 2-dimensional electrophoretic gels. *Biometrics,* **48**, 1273–1287.

Crawford, J. W., Ritz, K. and Young, I. M. (1993). Quantification of fungal morphology, gaseous transport and microbial dynamics in soil: an integrated framework using fractal geometry. *Geoderma,* **56**, 157–172.

Cressie, N. A. C. (1991). *Statistics for Spatial Data.* Wiley, New York.

Crimmins, T. R. (1982). A complete set of Fourier descriptors for two-dimensional shapes. *IEEE Transactions on Systems, Man and Cybernetics,* **12**, 848–855.

Danielsson, P. E. (1980). Euclidean distance mapping. *Computer Vision, Graphics and Image Processing,* **14**, 227–248.

Darbyshire, J. F., Griffiths, B. S., Davidson, M. S. and McHardy, W. J. (1989). Ciliate distribution amongst soil aggregates. *Revue d'Ecologie et de Biologie du Sol,* **26**, 47–56.

Davis, L. S. and Rosenfeld, A. (1978). Noise cleaning by iterative local averaging. *IEEE Transactions on Systems, Man and Cybernetics,* **8**, 705–710.

Devore, R. A., Jawerth, B. and Lucier, B. J. (1992). Image compression through wavelet transform coding. *IEEE Transactions on Information Theory,* **38**, 719–746.

Diggle, P. J. (1983). *Statistical Analysis of Spatial Point Patterns.* Academic Press, London.

Done, J. T., Upcott, D. H., Frewin, D. C. and Hebert, C. N. (1984). Atrophic Rhinitis: snout morphometry for quantitative assessment of conchal atrophy. *Veterinary Record,* **114**, 33–35.

Donoho, D. L., Johnstone, I. M., Hoch, J. C. and Stern, A. S. (1992). Maximum entropy and the nearly black object (with discussion). *Journal of the Royal Statistical Society, Series B,* **54**, 41–81.

Dorst, L. and Smeulders, A. W. M. (1987). Length estimators for digitized contours. *Computer Vision, Graphics and Image Processing,* **40**, 311–333.

Draper, N. H. and Smith, H. (1981). *Applied Regression Analysis.* Wiley, New York.

Durand, J. M., Gimonet, B. J. and Perbos, J. R. (1987). SAR data filtering for classification. *IEEE Transactions on Geoscience and Remote Sensing*, **25**, 629–637.

Eubank, R. L. (1988). *Spline Smoothing and Nonparametric Regression*. Marcel Dekker, New York.

Foley, J. D., Van Dam, A., Feiner, S. K. and Hughes, J. F. (1991). *Computer Graphics: Principles and Practice*. Addison-Wesley, Reading, Massachusetts.

Fong, Y., Pomalaza-Raez, C. A. and Wang, X. (1989). Comparison study of nonlinear filters in image processing applications. *Optical Engineering*, **28**, 749–760.

Fowler, P. A., Casey, C. E., Cameron, G. G., Foster, M. A. and Knight, C. H. (1990). Cyclic changes in composition and volume of the breast during the menstrual cycle, measured by magnetic resonance imaging. *British Journal of Obstetrics and Gynaecology*, **97**, 595–602.

Freeman, H. (1974). Computer processing of line-drawing images. *Computing Surveys*, **6**, 57–97.

Fu, K. S. (1974). *Syntactic Methods in Pattern Recognition*. Academic Press, New York.

Geman, S. and Geman, D. (1984). Stochastic relaxation, Gibbs distributions and the Bayesian restoration of images. *IEEE Transactions on Pattern Analysis and Machine Intelligence*, **6**, 721–735.

Glasbey, C. A. (1993). An analysis of histogram-based thresholding algorithms. *CVGIP: Graphical Models and Image Processing*, **55**, 532–537.

Glasbey, C. A., Abdalla, I. and Simm, G. (1994). Automatic interpretation of sheep ultrasound scans. (In preparation).

Glasbey, C. A., Hitchcock, D., Russel, A. J. F. and Redden, H. (1994). Towards automatic measurement of cashmere fibre diameter by image analysis. *Journal of the Textile Institute*, (in press).

Glasbey, C. A., Horgan, G. W. and Darbyshire, J. F. (1991). Image analysis and three-dimensional modelling of pores in soil aggregates. *Journal of Soil Science*, **42**, 479–486.

Glasbey, C. A., Horgan, G. W., Gibson, G. J. and Hitchcock, D. (1994). Fish shape analysis using landmarks. *Biometrical Journal* (in press).

Glasbey, C. A., Horgan, G. W. and Hitchcock, D. (1994). A note on the greyscale response and sampling properties of a desktop scanner. *Pattern Recognition Letters*, **15**, 705–711.

Glasbey, C. A., McRae, D. C. and Fleming, J. (1988). The size distribution of potato tubers and its application to grading schemes. *Annals of Applied Biology*, **113**, 579–587.

Glasbey, C. A. and Wright, F. G. (1994). An algorithm for unwarping multitrack electrophoretic gels. *Electrophoresis*, **15**, 143–148.

Gonzalez, R. C. and Wintz, P. (1987). *Digital Image Processing*, 2nd edition. Addison-Wesley, Reading, Massachusetts.

Goodall, C. (1991). Procrustes methods in the statistical analysis of shape (with discussion). *Journal of the Royal Statistical Society, Series B*, **53**, 285–339.

Gordon, A. D. (1981). *Classification*. Chapman and Hall, London.

Granlund, G. H. (1972). Fourier preprocessing for hand print character recognition. *IEEE Transactions on Computers*, **21**, 195–201.

Green, A. A., Berman, M., Switzer, P. and Craig, M. D. (1988). A transformation for ordering multispectral data in terms of image quality with implications for noise removal. *IEEE Transactions on Geoscience and Remote Sensing*, **26**, 65–74.

Grenander, U., Chow, Y. and Keenan, D. M. (1991). *Hands: A Pattern Theoretic Study of Biological Shapes*. Springer-Verlag, New York.

Griffin, H. G. and Griffin, A. M. (1993). DNA sequencing—recent innovations and future trends. *Applied Biochemistry and Biotechnology*, **38**, 147–159.

Haralick, R. M. (1983). Ridges and valleys on digital images. *Computer Vision, Graphics and Image Processing*, **22**, 28–38.

Haralick, R. M. (1984). Digital step edges from zero crossing of second directional derivatives. *IEEE Transactions on Pattern Analysis and Machine Intelligence*, **6**, 58–68.

Haralick, R. M. and Shapiro, L. G. (1985). Image segmentation techniques. *Computer Vision, Graphics and Image Processing*, **29**, 100–132.

Haralick, R. M. and Shapiro, L. G. (1992). *Computer and Robot Vision*. Addison-Wesley, Reading, Massachusetts.

Haralick, R. M., Sternberg, S. R. and Zhuang, X. (1987). Image analysis using mathematical morphology. *IEEE Transactions on Pattern Analysis and Machine Intelligence*, **9**, 532–550.

Haralick, R. M., Watson, L. T. and Laffey, T. J. (1983). The topographic primal sketch. *International Journal of Robotics Research*, **2**, 50–72.

Harwood, D., Subbarao, M., Hakalahti, H. and Davis, L. S. (1987). A new class of edge-preserving smoothing filters. *Pattern Recognition Letters*, **6**, 155–162.

Hastie, T. J. and Tibshirani, R. J. (1990). *Generalized Additive Models*. Chapman and Hall, London.

Heckbert, P. (1982). Color image quantization for frame buffer display. *SIGGRAPH '82*, 297–304.

Herbin, M., Venot, A., Devaux, J. Y., Walter, E., Lebruchec, J. F., Dubertret, L. and Roucayrol, J. C. (1989). Automated registration of dissimilar images—application to medical imagery. *Computer Vision, Graphics and Image Processing*, **47**, 77–88.

Hill, A. and Taylor, C. J. (1992). Model-based image interpretation using genetic algorithms. *Image and Vision Computing*, **10**, 295–300.

Holmes, T. J. and Levy, W. J. (1987). Signal-processing characteristics of differential-interference-contrast microscopy. *Applied Optics*, **26**, 3929–3939.

Horgan, G. W. (1994). Choosing weight functions for filtering SAR. *International Journal of Remote Sensing*, **15**, 1053–1064.

Horgan, G. W. and Ball, B. C. (1995). Simulating diffusion in a Boolean model of soil pores. (In press).

Horgan, G. W., Creasey, A. M. and Fenton, B. (1992). Superimposing two-dimensional gels to study genetic variation in malaria parasites. *Electrophoresis*, **13**, 871–875.

Horowitz, S. L. and Pavlidis, T. (1976). Picture segmentation by a tree traversal algorithm. *Journal of the Association for Computing Machinery*, **23**, 368–388.

Houle, M. E. and Toussaint, G. T. (1988). Computing the width of a set. *IEEE Transactions on Pattern Analysis and Machine Intelligence*, **10**, 761–765.

Hu, M. K. (1962). Visual pattern recognition by invariant moments. *IRE Transactions on Information Theory*, **8**, 179–187.

Huang, T. S., Yang, G. J. and Tang, G. Y. (1979). A fast two-dimensional median filtering algorithm. *IEEE Transactions on Acoustics, Speech and Signal Processing*, **27**, 13–18.

Imme, M. (1991). A noise peak elimination filter. *CVGIP: Graphical Models and Image Processing*, **53**, 204–211.

Jain, A. K. (1989). *Fundamentals of Digital Image Processing*. Prentice-Hall, Englewood Cliffs, NJ.

Jang, B. K. and Chin R. T. (1990). Analysis of thinning algorithms using mathematical morphology. *IEEE Transactions on Pattern Analysis and Machine Intelligence*, **12**, 541–551.

Jensen, J. R. (1986). *Introductory Digital Image Processing: A Remote Sensing Perspective*. Prentice-Hall, Englewood Cliffs, NJ.

Jeulin, D. (1993). Random models for morphological analysis of powders. *Journal of Microscopy*, **172**, 13–21.

Jones, M. C. and Sibson, R. (1987). What is projection pursuit? (with discussion). *Journal of the Royal Statistical Society, Series A*, **150**, 1–36.

Juhola, M., Katajainen, J. and Raita, T. (1991). Comparison of algorithms for standard median filtering. *IEEE Transactions on Signal Processing*, **39**, 204–208.

Kalles, D. and Morris, D. T. (1993). A novel fast and reliable thinning algorithm. *Image and Vision Computing*, **11**, 588–603.

Kass, M., Witkin, A. and Terzopoulos, D. (1988). Snakes: active contour models. *International Journal of Computer Vision*, **1**, 321–331.

Khurshid, F. and Beck, S. (1993). Error analysis in manual and automated DNA sequencing. *Analytical Biochemistry*, **208**, 138–143.

Kirby, R. L. and Rosenfeld, A. (1979). A note on the use of (gray level, local average gray level) space as an aid in threshold selection. *IEEE Transactions on Systems, Man and Cybernetics*, **9**, 860–864.

Kittler, J. and Illingworth, J. (1986). Minimum error thresholding. *Pattern Recognition*, **19**, 41–47.

Koplowitz, J. and Bruckstein, A. M. (1989). Design of perimeter estimators for digitized planar shapes. *IEEE Transactions on Pattern Analysis and Machine Intelligence*, **11**, 611–622.

Krzanowski, W. J. (1988). *Principles of Multivariate Analysis : A User's Perspective*. Clarendon Press, Oxford.

Leavers, V. F. (1992). *Shape Detection in Computer Vision Using the Hough Transform*. Springer-Verlag, London.

Lee, J. S. (1981). Refined filtering of image noise using local statistics. *Computer Graphics and Image Processing*, **15**, 380–389.

Lee, J. S. (1983). A simple speckle smoothing algorithm for synthetic aperture radar images. *IEEE Transactions on Systems, Man and Cybernetics*, **13**, 85–89.

Lempel, A. and Ziv, J. (1986). Compression of 2-dimensional data. *IEEE Transactions on Information Theory*, **32**, 2–8.

Lev, A., Zucker, S. W. and Rosenfeld, A. (1977). Iterative enhancement of noisy images. *IEEE Transactions on Systems, Man and Cybernetics*, **7**, 435–442.

Mallat, S. G. (1989). A theory for multiresolution signal decomposition: the wavelet representation. *IEEE Transactions on Pattern Analysis and Machine Intelligence*, **11**, 674–693.

Maltin, C. A., Hay, S. M., Delday, M. I., Lobley, G. E. and Reeds, P. J. (1989). The action of the β-agonist clenbuterol on protein metabolism in innervated and denervated phasic muscles. *Biochemistry Journal*, **261**, 965–971.

Mandelbrot, B. B. (1982). *The Fractal Geometry of Nature*. Freeman, San Francisco.

Mardia, K. V. and Hainsworth, T. J. (1988). A spatial thresholding method for image segmentation. *IEEE Transactions on Pattern Analysis and Machine Intelligence*, **10**, 919–927.

Mardia, K. V. and Hainsworth, T. J. (1989). Statistical aspects of moment invariants in image analysis. *Journal of Applied Statistics*, **16**, 423–435.

Maronna, R. and Jacovkis, P. M. (1974). Multivariate clustering procedures with variable metrics. *Biometrics*, **30**, 499–505.

Marr, D. and Hildreth, E. (1980). Theory of edge detection. *Proceedings of the Royal Society, London, Series B*, **207**, 187–217.

Marshall, S. (1989). Review of shape coding techniques. *Image and Vision Computing*, **7**, 281–294.

Martelli, A. (1976). An application of heuristic search methods to edge and contour detection. *Communications of the Association for Computing Machinery*, **19**, 73–83.

Martin, N. J. and Fallowfield, H. J. (1989). Computer modelling of algal waste treatment systems. *Water Science and Technology*, **21**, 277–287.

Martinelli, G., Ricotti, L. P. and Marcone, G. (1993). Neural clustering for optimal KLT image compression. *IEEE Transactions on Signal Processing*, **41**, 1737–1739.

Mastin, G. A. (1985). Adaptive filters for digital image noise smoothing: an evaluation. *Computer Vision, Graphics and Image Processing*, **31**, 103–121.

Meyer, F. (1989). Skeletons and perceptual graphs. *Signal Processing*, **16**, 335–363.

Meyer, F. (1992). Mathematical morphology: from two dimensions to three dimensions. *Journal of Microscopy*, **165**, 5–28.

Meyer, F. and Beucher, S. (1990). Morphological segmentation. *Journal of Visual Communication and Image Representation*, **1**, 21–46.

Mitchell, H. B. and Mashkit, N. (1992). Noise smoothing by a fast k-nearest neighbour algorithm. *Signal Processing: Image Communication*, **4**, 227–232.

Mou, D. and Stoermer, E. F. (1992). Separating *Tabellaria* (Bacillariophyceae) shape groups based on Fourier descriptors. *Journal of Phycology*, **28**, 386–395.

NAG (1991). *Numerical Algorithms Group. Library Manual Mark 15*. NAG Central Office, 256 Banbury Road, Oxford OX2 7DE, UK.

Nagao, M. and Matsuyama, T. (1979). Edge preserving smoothing. *Computer Graphics and Image Processing*, **9**, 394–407.

Nasiopoulos, P., Ward, K. and Morse, D. J. (1991). Adaptive compression coding. *IEEE Transactions on Communications*, **39**, 245–1254.

Nason, G. P. and Sibson, R. (1991). Using projection pursuit in multispectral image analysis. *Proceedings of the 23rd Symposium on the Interface: Computer Science and Statistics*, (ed. E.M. Keramidas), 579–582.

Nokihara, K., Morita, N. and Kuriki, T. (1992). Applications of an automated apparatus for 2-dimensional electrophoresis, model TEP-1, for microsequence analyses of proteins. *Electrophoresis*, **13**, 701–707.

Oliver, C. J. (1991), Information from SAR images. *Journal of Physics D: Applied Physics*, **24**, 1493–1514.

Pavlidis, T. (1977). Polygonal approximations by Newton's method. *IEEE Transactions on Computers*, **26**, 800–807.

Pavlidis, T. (1978). A review of algorithms for shape analysis. *Computer Graphics and Image Processing*, **7**, 243–258.

Pavlidis, T. (1980). Algorithms for shape analysis of contours and waveforms. *IEEE Transactions on Pattern Analysis and Machine Intelligence*, **4**, 301–312.

Pavlidis, T. and Liow, Y. (1990). Integrating region growing and edge detection. *IEEE Transactions on Pattern Analysis and Machine Intelligence*, **12**, 225–233.

Press, W. H., Flannery, B. P., Teukolsky, S. A. and Veltering, W. T. (1992) *Numerical Recipes in FORTRAN: The Art of Scientific Computing*, 2nd edition. Cambridge University Press, Cambridge.

Purll, D. J. (1985). Solid-state image sensors. *Automated Visual Inspection*, (ed. B. G. Batchelor, D. A. Hill and D. C. Hodgson). IFS (Publications) Ltd, Bedford, 255–293.

Qian, W. and Titterington, D. M. (1991). Pixel labelling for 3-dimensional scenes based on Markov mesh models. *Signal Processing*, **22**, 313–328.

Reed, T. R. and du Buf, J. M. H. (1993). A review of recent texture segmentation and feature-extraction techniques. *CVGIP: Image Understanding*, **57**, 359–372.

Reiss, T. H. (1991). The revised fundamental theorem of moment invariants. *IEEE Transactions on Pattern Analysis and Machine Intelligence*, **13**, 830–834.

Ridler, T. and Calvard, S. (1978). Picture thresholding using an iterative selection method. *IEEE Transactions on Systems, Man and Cybernetics*, **8**, 630–632.

Rignot, E. J. M., Kowk, R., Curlander, J. C. and Pang, S. S. (1991). Automated multi-sensor registration—requirements and techniques. *Photogrammetric Engineering and Remote Sensing*, **57**, 1029–1038.

Ripley, B. D. and Sutherland, A. I. (1990). Finding spiral structures in images of galaxies. *Philosophical Transactions of the Royal Society, London, Series A*, **332**, 477–485.

Ritz, K. and Crawford, J. W. (1990). Quantification of the fractal nature of colonies of *Trichoderma viride*. *Mycological Research*, **94**, 1138–1141.

Robertson, J. F., Wilson, D. and Smith, W. J. (1990). Atrophic rhinitis: the influence of the aerial environment. *Animal Production*, **50**, 173–182.

Rohlf, F. J. and Archie, J. W. (1984). A comparison of Fourier methods for the description of wing shape in mosquitoes (Diptera: Culicidae). *Systematic Zoology*, **33**, 302–317.

Rosenfeld, A. (1974). Compact figures in digital pictures. *IEEE Transactions on Systems, Man and Cybernetics*, **4**, 211–223.

Rosenfeld, A. and Kak, A. C. (1982). *Digital Picture Processing*, 2nd edition. Academic Press, San Diego.

Rosenfeld, A. and Pfaltz, J. L. (1966). Sequential operations in digital picture processing. *Journal of the Association for Computing Machinery*, **13**, 471–479.

Russel, A. J. F. (1991). Cashmere production—the viable alternative. *Outlook on Agriculture*, **20**, 39–43.

Samadani, R. and Han, C. (1993). Computer-assisted extraction of boundaries from images. *SPIE Storage and Retrieval for Image and Video Databases, San Jose, USA, February 1993; SPIE Proceedings*, **1908**, 219–225.

Serra, J. (1982). *Image Analysis and Mathematical Morphology*. Academic Press, London.

Serra, J. (1986). Introduction to mathematical morphology. *Computer Vision, Graphics and Image Processing*, **35**, 283–305.

Serra, J. (ed.) (1988). *Image Analysis and Mathematical Morphology*. Volume 2: *Theoretical Advances*. Academic Press, London.

Shaw, P. J. and Rawlins, D. J. (1991). Three-dimensional fluorescence microscopy. *Progress in Biophysics and Molecular Biology*, **56**, 187–213.

Silverman, B. W., Jones, M. C., Wilson, J. D. and Nychka, D. W. (1990). A smoothed EM approach to indirect estimation problems, with particular reference to stereology and emission tomography (with discussion). *Journal of the Royal Statistical Society, Series B*, **52**, 271–324.

Simm, G. (1992). Selection for lean meat production in sheep. *Recent Advances in Sheep and Goat Research* (ed. A.W. Speedy). CAB International, 193–215.

Skilling, J. and Bryan, R. K. (1984). Maximum entropy image reconstruction: general algorithm. *Monthly Notices of the Royal Astronomical Society*, **211**, 111–124.

Skolnick, M. M. (1986). Application of morphological transformations to the analysis of two-dimensional electrophoretic gels of biological materials. *Computer Vision, Graphics and Image Processing*, **35**, 306–332.

Skolnik, M. I. (1981). *Introduction to Radar Systems*, 2nd edition. McGraw-Hill, New York.

Smith, L. M. (1993). Automated DNA-sequencing—a look into the future. *Cancer Detection and Prevention*, **17**, 283–288.

Solomon, J. E. and Harrington, M. G. (1993). A robust, high-sensitivity algorithm for automated detection of proteins in 2-dimensional electrophoresis gels. *Computer Applications in the Biosciences*, **9**, 133–139.

Stark, H. (1987). *Image Recovery: Theory and Applications*. Academic Press, New York.

Sternberg, S. R. (1986). Greyscale morphology. *Computer Vision, Graphics and Image Processing*, **35**, 333–355.

Stoyan, D. (1990). Stereology and stochastic geometry. *International Statistical Review*, **58**, 227–242.

Stoyan, D., Kendall, W. S. and Mecke, J. (1987). *Stochastic Geometry and its Applications*. Wiley, Chichester.

Strachan, N. J. C. (1993). Recognition of fish species by color and shape. *Image and Vision Computing*, **11**, 2–10.

Strachan, N. J. C., Nesvadba, P. and Allen, A.R. (1990a). Fish species recognition by shape analysis of images. *Pattern Recognition*, **5**, 539–544.

Strachan, N. J. C., Nesvadba, P. and Allen, A. R. (1990b). Calibration of a video camera digitising system in the CIE $L^*u^*v^*$ colour space. *Pattern Recognition Letters*, **11**, 771–777.

Tailor, A., Cross, A., Hogg, D. C. and Mason, D. C. (1986). Knowledge-based interpretation of remotely sensed images. *Image and Vision Computing*, **4**, 67–83.

Teh, C. H. and Chin, R. T. (1986). On digital approximation of moment invariants. *Computer Vision, Graphics and Image Processing*, **33**, 318–326.

Teh, C. H. and Chin, R. T. (1988). On image analysis by the methods of moments. *IEEE Transactions on Pattern Analysis and Machine Intelligence*, **10**, 496–513.

Tomita, F. and Tsuji, S. (1977). Extraction of multiple regions by smoothing in selected neighborhoods. *IEEE Transactions on Systems, Man and Cybernetics*, **7**, 107–109.

Ton, J., Sticklen, J. and Jain, A. K. (1991). Knowledge-based segmentation of Landsat images. *IEEE Transactions on Geoscience and Remote Sensing*, **29**, 222–231.

Torre, V. and Poggio, T. A. (1986). On edge detection. *IEEE Transactions on Pattern Analysis and Machine Intelligence*, **2**, 147–163.

Trubuil, A. (1993). Analysis of one-dimensional electrophoregrams. *Computer Applications in the Biosciences*, **9**, 451–458.

Trussell, H. J. (1979). Comments on 'Picture thresholding using an iterative selection method'. *IEEE Transactions on Systems, Man and Cybernetics*, **9**, 311.

van Herk, M. (1992). A fast algorithm for local minimum and maximum filters on rectangular and octagonal kernels. *Pattern Recognition Letters*, **13**, 517–521.

Vilenkin, N. Y. (1968). *Stories about Sets.* Academic Press, New York.

Vincent, L. (1993). Morphological greyscale reconstruction in image analysis: applications and efficient algorithms. *IEEE Transactions on Image Processing*, **2**, 176–201.

Vincent, L. and Soille, P. (1991). Watersheds in digital spaces: an efficient algorithm based on immersion simulations. *IEEE Transactions on Pattern Analysis and Machine Intelligence*, **13**, 583–598.

Wallace, G. (1991). The JPEG still picture compression standard. *Communications of the Association for Computing Machinery*, **34**, 30–44.

Wang, D. C. C., Vagnucci, A. H. and Li, C. C. (1981). Image enhancement by gradient inverse weighted smoothing scheme. *Computer Graphics and Image Processing*, **15**, 167–181.

Wang, D. C. C., Vagnucci, A. H. and Li, C. C. (1983). Digital image enhancement: a survey. *Computer Vision, Graphics and Image Processing*, **24**, 363–381.

Wells, W. M. (1986). Efficient synthesis of Gaussian filters by cascaded uniform filters. *IEEE Transactions on Pattern Analysis and Machine Intelligence*, **8**, 234–239.

Weszka, J. S. and Rosenfeld, A. (1979). Histogram modification for threshold selection. *IEEE Transactions on Systems, Man and Cybernetics*, **9**, 38–52.

Wu, W., Wang, M. J. and Liu, C. (1992). Performance evaluation of some noise reduction methods. *CVGIP: Graphical Models and Image Processing*, **54**, 134–146.

Xia, Y. (1989). Skeletonization via the realization of the fire front's propagation and extinction in digital binary shapes. *IEEE Transactions on Pattern Analysis and Machine Intelligence*, **11**, 1076–1086.

Young, I. M. and Crawford, J. W. (1991). The fractal structure of soil aggregates—its measurement and interpretation. *Journal of Soil Science*, **42**, 187–192.

Zahn, C. T. and Roskies, R. Z. (1972). Fourier descriptors for plane closed curves. *IEEE Transactions on Computers*, **21**, 269–281.

Author Index

Index compiled by Geoffrey C. Jones

Subject Index

Index compiled by Geoffrey C. Jones